Ticks of Trinidad and Tobago—An Overview

Ticks of Trinidad and Tobago—An Overview

Asoke K. Basu, MVSc, PhD
Professor in Veterinary Parasitology School of Veterinary Medicine,
Faculty of Medical Sciences, The University of the West Indies,
St. Augustine, Trinidad and Tobago

Roxanne A. Charles, DVM, MSc
Lecturer in Veterinary Parasitology School of Veterinary Medicine,
Faculty of Medical Sciences, The University of the West Indies,
St. Augustine, Trinidad and Tobago

ACADEMIC PRESS

An imprint of Elsevier
elsevier.com

Academic Press is an imprint of Elsevier
125 London Wall, London EC2Y 5AS, United Kingdom
525 B Street, Suite 1800, San Diego, CA 92101-4495, United States
50 Hampshire Street, 5th Floor, Cambridge, MA 02139, United States
The Boulevard, Langford Lane, Kidlington, Oxford OX5 1GB, United Kingdom

British Library Cataloguing-in-Publication Data
A catalogue record for this book is available from the British Library

Library of Congress Cataloging-in-Publication Data
A catalog record for this book is available from the Library of Congress

ISBN: 978-0-12-809744-1

For Information on all Academic Press publications
visit our website at https://www.elsevier.com

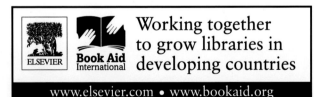

Working together
to grow libraries in
developing countries

www.elsevier.com • www.bookaid.org

Publisher: Sara Tenney
Acquisition Editor: Linda Versteeg-Buschman
Editorial Project Manager: Joslyn Chaiprasert-Paguio
Production Project Manager: Stalin Viswanathan

Typeset by MPS Limited, Chennai, India

CONTENTS

FOREWORD

It is indeed a great privilege and honor for me as Pro Vice-Chancellor and Campus Principal of The University of the West Indies (UWI) St. Augustine campus to write the foreword for this book as I continue to encourage academic excellence among the faculty and staff of our university. The UWI supports research and innovation by awarding grants to academic and research members of staff. A substantial amount of money is spent every year for this purpose as our university views quality of its academic output and service as a central pillar of its work.

This book, *Ticks of Trinidad and Tobago: An Overview*, is the first authoritative book on the ticks of this region. I consider this book, therefore, to possess significant academic and practical value. Indeed, ticks and tick-borne diseases are of economic importance worldwide because of their veterinary and public health significance. In fact, the prevalence and/or a lack of control of ticks in livestock and humans are responsible for billions of dollars in losses worldwide.

The authors have provided an in-depth review of all the available literature on ticks in Trinidad and Tobago for over 10 decades. Although ticks of Trinidad and Tobago are the focus of this book, general information on the morphology, identification, life cycle, and reproduction of ticks are included for the benefit of readers. This book will therefore serve as a useful guide to veterinary students, practitioners, researchers, zoologists, and lecturers of parasitology, not only in this region but also in other tropical areas. The book will be of great value in the diagnosis and control of ticks and tick-borne diseases.

The authors are highly qualified and experienced parasitologists. Professor A. Basu has over 30 years of teaching and research experience in Veterinary Parasitology in India, Africa, and the West Indies and has brought all this experience to bear in this book. I use this opportunity to congratulate the authors for an outstanding publication and a job well done. I am always delighted when my academic collea-gues of UWI do nationally and regionally impacting work and share their knowledge and expertise with a wide audience. This publication therefore merits our accolades.

Professor Clement K. Sankat
Pro Vice-Chancellor and Campus Principal
The University of the West Indies, St. Augustine,
Trinidad and Tobago

PREFACE

Ticks are very important parasites in animals and humans due to their blood-feeding activity and ability to transmit a number of pathogenic diseases to their affected hosts. These ectoparasites can be found in both tropical and temperate climates with preference for the former.

This book is intended to guide you, the reader (veterinary student, researcher, technician, veterinarian, biologist, ecologist, and persons in the medical field) on the tick species found in Trinidad and Tobago and in some cases, other parts of the Caribbean region as well as North, Central, and South America. This guide begins with a general account of hard and soft ticks including characteristic features of the taxon, their importance, general morphology, habits and general life cycles, methods of control, and their collection and preservation. The rest of this book deals with the 23 recorded tick species of Trinidad and Tobago. For each species mentioned, key aspects of their identification (including electron micrographs and line drawings), distribution, hosts, and disease transmission are emphasized.

It is the hope of the authors that this book be the "go to" guide for anyone wishing to identify ticks found on both islands.

ACKNOWLEDGMENTS

We are indebted to the authors of the book *Ticks of Domestic Animals in Africa: A Guide to Identification of Species*, which inspired us to compile all documents on tick species of Trinidad and Tobago and write this overview.

We would like to express our deep gratitude to Professor A.R. Walker and Dr. A. Guglielmone for encouragement and providing literatures, pictures, and other documents whenever required.

We are grateful to Dr. O. Voltzit, Zoological Museum, Moscow, Professor Serge Kraiter, Editor in Chief, Acarologia and Miss Louise Eyre of Oxford University Press for consenting to use some pictures from their publications/journals. We are very thankful to Andre G. Wolff, Linda Versteeg-Buschman and Halima N. Williams of the Elsevier for their constant help during the preparation of the manuscript.

We also extend our gratitude to T.H.G Aitken, G.M. Kohls, and E.S. Tikasingh, the pioneer researchers on ticks in Trinidad and Tobago for their collection of ticks during 1957−70, which have been preserved and kept in the Zoology Museum, Department of Life Sciences, the University of the West Indies, St. Augustine. We are thankful to Mr. Mike G. Rutherford, the Curator of the same museum, for providing us with the preserved ticks for photography.

Our sincere thanks to Mr. David Hinds of the Electron Microscopy Unit, Faculty of Medical Sciences, the University of the West Indies (UWI), Trinidad and Tobago for his expertise in preparation of the tick specimens for electron microscopy and photography. We are also very thankful to Mr. Neil Anderson Joseph of the Centre for Medical Sciences Education (CMSE), the UWI, for editing the photographs and Mr. Sukanta Roy, Bhairab Ganguly College, Kolkata, for assistance with editing of the references. We also appreciate the office support extended by Soumalya Sinha and Deepmalya Sinha.

We must express our special appreciation to Dr. Bidyadhar Sa of the CMSE, Dr. Karla Georges, and Professor Andrew Adogwa of the UWI School of Veterinary Medicine for their invaluable support and comments.

We also deeply appreciate the constant support and encouragement from Alden and Harmony while writing this book.

Last but foremost, we sincerely acknowledge Dr. Mala Basu, who has been with us assisting in all stages during the course of writing this book.

Cellular organisms are divided in two groups known as prokaryotes and eukaryotes. The genetic material, deoxyribonucleic acid (DNA) and ribonucleic acid (RNA) are arranged into structures called chromosomes. In the eukaryotic cell, the chromosomes are surrounded by a nuclear membrane forming a true nucleus. Eukaryotic cells contain a variety of organelles and a cytoskeleton, which is composed of microtubules, microfilaments, and intermediate filaments. The general structures of prokaryotic cells are a plasma membrane, cytoplasm, ribosomes, and genetic material in the form of DNA and RNA. Unlike eukaryotes, the genetic material in prokaryotes is not surrounded by a nuclear membrane and lies in a region called the nucleoid. A prokaryotic chromosome is circular but in eukaryotes, the chromosome is linear. Some cells have other structures such as cell wall, pili, and flagella. The cell components play a vital role in survival, growth, and reproduction of the cell.

The phylum Arthropoda (arthropods) belongs to the kingdom Animalia and super kingdom Eukaryota. Arthropods (from Greek arthro-, joint + podos, foot) are invertebrates, which possess a chitinous exoskeleton, jointed appendages, and a segmented body. They include insects, arachnids, crustaceans, and myriapods. This is the largest phylum and consists of the most successful animals on the planet. They have adapted successfully to life in water, on land and in the air.

Ticks belong to the phylum Arthropoda, Class Arachnida, Order Acari, and Suborder Ixodida (Fig. 1.1). The Suborder Ixodida consists of two families: the family Ixodidae (the hard ticks) with marked sexual dimorphism, and the family Argasidae (the soft ticks) in which dimorphism is not distinct.

Ticks feed on the blood of mammals (including humans), birds, reptiles, and amphibians. When ticks parasitize large mammals, they are called macromastophiles and on small mammals such as rodents, they are called micromastophiles. Ticks feeding on birds are called ornithophiles and those on reptiles and amphibians are known as herpetophiles.

The total number of valid tick species recorded from all climatic zones throughout the world is 896 (702 Ixodid, 193 Argasid, and 1 Nuttalliella tick (sub-) species).[1]

Ticks are of considerable importance in the veterinary and medical field[2] since some ticks are vectors of numerous pathogens.[3] Ranking second to mosquitoes as vectors of human infectious diseases, ticks are known to transmit arboviruses (e.g., tick-borne encephalitis virus and other Flaviviridae, several Reoviridae, Bunyaviridae, and Iridoviridae), bacteria (*Rickettsia*, *Ehrlichia*, *Borrelia*), and protists (*Babesia* and *Theileria*).[4,5] Nijhof et al.[6] recorded the vector competence of ticks, including 808 tick−pathogen relationships: 322 relationships with 84 bacteria, 302 with 124 viruses, 143 with 59 apicomplexan parasites, 4 with 3 nematodes and 3 with *Trypanosoma theileri*, and an additional 34 species of ticks are found associated with toxicosis. These authors also recorded 233 reported cases of acaricide resistance for 20 tick species of veterinary and medical importance. About 80% of the world's cattle are infested with ticks.[7,8] Globally, the annual cost associated with ticks and tick-borne diseases in cattle alone is estimated to be in the billions of US dollars.[9]

Ticks also cause annoyance and direct blood loss to their hosts due to their blood-sucking habits. A single adult female can suck 0.5−2.0 mL of blood and several thousand ticks can cause daily blood loss of several hundred milliliters.[10] A fully engorged female hard tick can store large volumes of blood increasing her weight 200−600 times her unfed weight. Conversely, female soft ticks have shown to ingest lesser volumes of host blood than female hard ticks, only 5−10 times of their unfed body weight.[11] Ticks, by their biting and blood-sucking habits, may cause anemia, restlessness and tick-worry, loss of condition, decrease in milk production, debilitation, and tick paralysis of their hosts. Tick bites also cause skin damage thereby reducing the value of the animal hides. Gross lesions due to tick bites result in focal erosions, erythema, crusted ulcers with alopecia and nodules in some host animals.

The toxin liberated into the host's blood stream by some ticks during blood feeding may cause a variety of toxicoses including sweating sickness and tick paralysis. Paralysis may occur due to toxins in the tick's salivary secretions when the point of attachment is near a vital nerve. The degree of paralysis is proportional to the length of time the tick has been feeding and, frequently, on the number of ticks attached.[10]

Cattle and buffalo are especially susceptible to tick infestation, which is a great menace in Trinidad and Tobago. Some common tick-borne diseases endemic in these ruminants are babesiosis and anaplasmosis.[12]

A number of researchers have reported on several aspects of problems caused by ticks in Trinidad and Tobago.[13–25] Others reported on the incidence and taxonomy, distribution, biology and bionomics of ticks.[24,26] Aitken et al.,[27] Lans,[28] Polar et al.,[29] and Polar[30] worked on the treatment against tick infestation and Aitken et al.,[27] Asgarali et al.,[31] and Georges[32] studied their disease transmission. However, there is inadequate information on ticks in Trinidad and Tobago thus warranting further studies on tick biology, bionomics, pathogenesis, disease transmission, treatment, and control measures.

CHAPTER 1

A General Account of Ticks

SYSTEMATIC POSITION

The systematic position and classification of ticks in relation to other arthropods are shown in Fig. 1.1.

CHARACTERISTIC FEATURES OF THE TAXA

Phylum Arthropoda

- Bilaterally symmetrical body
- Body segmented in regions called tagmata—head, thorax, and abdomen
- Jointed, paired appendages (e.g., legs and antennae)
- Body covered with chitinous exoskeleton that shed during growth
- Well-developed digestive system
- Open circulatory system present
- Ocelli and compound eyes present
- Sexes separate with well-developed gonads

Class Arachnida

- Members are largely carnivorous and terrestrial
- Adults have four pairs of legs, with the exception of a few mites
- Two additional pairs of appendages are the chelicerae used for feeding or defense, and the pedipalps used for feeding, locomotion, and/or reproduction
- Antennae and wings are absent
- Body divided into two segments—cephalothorax (fusion of head and thorax) and abdomen or gnathosoma (head) and idiosoma (fusion of thorax and abdomen)

The arachnids are most commonly confused with insects. The differences between the Class Insecta and Arachnida are shown in Table 1.1.

Ticks of Trinidad and Tobago—An Overview. DOI: http://dx.doi.org/10.1016/B978-0-12-809744-1.00001-3

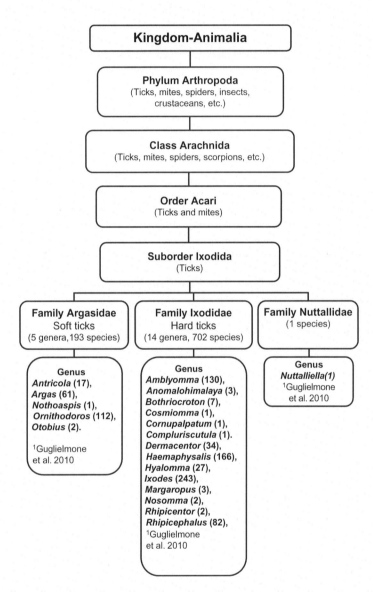

Figure 1.1 Systematic position and classification of ticks in relation to other arthropods.

Order Acari (Acarina)

- Nymphs and adults possess four pairs of legs, larvae have only three pairs
- Members of some families have a dense covering of setae, absent in others

Table 1.1 Differences Between Insects and Arachnids		
Characters	**Class Insecta**	**Class Arachnida**
Body segments	Head Thorax Abdomen	Gnathosoma (head) and idiosoma (fused thorax and abdomen) or cephalothorax (fused head and thorax) and abdomen
Number of appendages	Three pairs	Adults have four pairs
Antennae	Single pair of antennae	No antennae
Eyes	Compound eyes	Simple eyes
Wings	Present in many	Absent
Jaws	Mandibles present	No mandible, chelicerae present
Metamorphosis	Occurs in many	No metamorphosis
Examples	Fly Flea	Tick Mite

- The body is strongly sclerotized with numerous hardened plates while the soft-bodied mites and ticks have a few hardened plates or shields

This is the only group of major veterinary importance under the Class Arachnida. The mites and ticks belong to this order.

Suborder Ixodida
- All ticks belong to this suborder
- They are blood-sucking obligatory ectoparasites
- Mouthparts with hypostome possess recurved teeth for anchoring on the host
- Haller's organ with olfactory function present on first tarsi
- The suborder Ixodida is divided into three families:
  - Family Argasidae (Soft ticks)
  - Family Ixodidae (Hard ticks)
  - Family Nuttalliellidae

Family Argasidae
- Members known as soft ticks
- The integument is leather-like
- No dorsal slit or scutum
- Capitulum and mouthparts subterminal and not visible from the dorsal aspect
- One pair of spiracles situated between the coxae of legs III and IV
- Palpi freely articulated and leg-like
- No porose area
- Coxa unarmed
- Festoons absent
- Eyes, if present, are situated on the lateral sides on supra-coxal folds

Family Ixodidae
- Members known as hard ticks
- The integument is hard
- Chitinous shield or scutum present
- In the case of males, the scutum extends over the entire dorsal surface; in female it extends only to a small portion behind the head
- The mouthpart is situated anteriorly and is well visible from the dorsal aspect
- The spiracle is situated behind the coxa of leg IV
- Palpi rigid
- Porose areas present
- Festoons generally present
- Eyes situated on the side of scutum
- Coxae generally armed with spurs
- Tarsi generally armed with one or two ventral parts
- Pulvilli always present
- Camerostome absent

Family Argasidae (Soft ticks) and Ixodidae (Hard ticks) are important in veterinary and human medicine. The differences between their characteristic features are shown in Table 1.2.

GENERAL IDENTIFICATION CHARACTERS OF THE TICK GENERA PRESENT IN TRINIDAD AND TOBAGO

Identification characters of ticks by family, genus and species are available from different parts of the world. Adult ticks are easier to

Table 1.2 Differences Between Soft and Hard Ticks

Soft Tick	Hard Tick
Family: Argasidae	Family: Ixodidae
1. The integument is leathery with no dorsal shield or scutum	1. The integument is hard and a chitinous shield or scutum is present
2. The mouthparts situated anteriorly on the ventral surface and not visible from the dorsal aspect	2. The mouthparts situated anteriorly and well visible from the dorsal aspect
3. A pair of spiracles situated between the coxae of legs III and IV	3. Pair of spiracles situated behind the coxa of leg IV
4. Palpi leg-like	4. Palpi rigid
5. No porose areas	5. Porose areas present
6. Festoons absent	6. Festoons generally present
7. Eyes on the lateral surface of the body on supra-coxal folds	7. Eyes on the side of scutum
8. Coxae unarmed	8. Coxae generally armed with spurs
9. Tarsi without ventral parts	9. Tarsi generally armed with one or two ventral parts
10. Pulvilli absent or rudimentary	10. Pulvilli always present
11. Camerostome present	11. Camerostome absent
12. Sexual dimorphism not distinct	12. Sexual dimorphism well marked
13. Osmoregulation through coxal glands	13. Osmoregulation through salivary glands
14. Ticks feed rapidly on hosts (for about an hour), then drop off the host	14. Ticks attach to their hosts for up to several days while feeding
15. Mating occurs off the hosts	15. Mating occurs on the hosts (except *Ixodes*)
16. Adult females take multiple blood meals and produce a small batch of eggs after each blood meal (generally 6−7 batches in its lifetime). Each small batch consists of 50−200 eggs	16. Fully engorged adult females drop down from the host, lay a large numbers of eggs (2000−18,000) and then die
17. Life cycle includes egg, larva, two or more nymphal stages, and adult	17. Life cycle includes egg, larva, nymph, and adult stage
18. Eggs are larger than those of ixodid ticks	18. Eggs are smaller than those of argasid ticks
19. All are multihost (except *Otobius megnini*, a one-host tick)	19. Either a one-, two-, or three-host tick depending on species
20. Nymphs need blood meal several times and go through several nymphal stages (generally 3−5)	20. Single stage of nymph transforms to adult after blood meal
21. Can live up to 10 years[4]	21. Life span ranges from ~2 months to ≥ 3 years[4]

differentiate than immature stages and hence the former is used for identification. In recent years, molecular methods have been developed to identify ticks, and it is expected that in the future such methods can be used for the differentiation of closely related species.[33]

The tick genera recorded from Trinidad and Tobago are *Argas*, *Ornithodoros*, *Amblyomma*, *Dermacentor*, *Haemaphysalis*, *Ixodes*, and *Rhipicephalus*. The identification characters of these seven genera are stated below.[34–36]

Genus *Argas*
- Margin of body with definite lateral suture
- Eyes absent
- Edge of body flattened

Genus *Ornithodoros*
- Margin of body thick, rounded, lateral suture absent
- Integument mamillated
- Hypostome with well-developed teeth
- Eyes absent

Genus *Amblyomma*
- Large ticks (unfed female up to 8mm long; fed female up to 20 mm long)
- Mouthparts (inflict deep painful bites) longer than capituli
- Eyes (flat) and festoons present
- Legs often long and banded
- Scutum often highly ornate
- Spiracle subtriangular or comma shaped
- Anal groove situated posterior to anus

Genus *Dermacentor*
- Basis capituli rectangular with straight lateral margins
- Hypostomal dentition: 4/4
- Eyes and festoons present
- Palpal articles II broad
- Coxa IV much larger than coxae I, II, & III
- Coxa of leg I has two large spurs of equal length
- Spiracle goblets scattered over the spiracle plates
- Ventral shields absent in males
- Anal groove absent

Genus *Haemaphysalis*
- Inornate
- Capitulum short, palpal segment II broad and spreads laterally
- Eyes absent

- Coxa of leg I has single internal spur
- Festoons present
- Anal groove posterior to anus

Genus *Ixodes*
- Inornate
- Mouthparts long
- Basis capituli with straight lateral margins
- Eyes and festoons absent
- Palpal articles II longer than articles I and III
- Striations on integument
- Legs slender
- Pulvilli present
- Lateral suture absent
- Anal groove distinct and situated anterior to the anus

Genus *Rhipicephalus*
- Inornate
- Hexagonal basis capituli
- Eyes and festoons present
- Legs slender
- Spiracular plates large
- Ventral plates present only in males
- Anal groove posterior to the anus

IMPORTANCE OF TICKS

Ticks are of veterinary and public health importance for a number of reasons. They may cause:

- Direct loss of blood by sucking (0.5–2 mL of blood sucked by a single female hard tick).
- Annoyance to the host due to irritation caused by biting.
- Tick worry—may be relating to the devastating effects of heavy tick infestations leading to serious loss of weight, anemia and local secondary bacterial infections due to tick bites.
- Tick paralysis—some ticks e.g., *Ixodes* and *Dermacentor* species liberate neurotoxins during blood feeding, which causes flaccid paralysis of the host's fore- and/or hindlimbs.

- Skin damage due to insertion of hypostome (mouthpart) into the host's skin.
- Expenses incurred for treatment with acaricides.
- Diseases transmission to domestic/wild animals and humans. Ticks rank second to mosquitoes as vectors of human diseases.

SITE OF ATTACHMENT ON HOSTS

Most ticks have a specific site of preference on the body of the host. Table 1.3 summarizes the site preferences exhibited by some tick species.

Table 1.3 Site Preferences Exhibited by Ticks on Their Hosts	
Tick Species	**Common Site of Attachment on the Body in Order of Preference**
Haemaphysalis bispinosa	Ear, limbs, dewlaps, neck, tail, axilla, groin, and abdomen
Rhipicephalus (Boophilus) microplus	Ear, limbs, dewlaps, abdomen, and chest
Rhipicephalus (Boophilus) decoloratus	Abdomen, limbs, dewlaps, and groin
Amblyomma variegatum	Under the tail, margin of the anus, limbs, and groin
Amblyomma cajennense	Skin, lower body surface, especially between the legs
Hyalomma anatolicum anatolicum	Chest, abdomen, neck, udder, and scrotum
Rhipicephalus evertsi evertsi	Nape of the neck, under the tail, and margin of the anus

COMMON NAMES OF TICKS

Genus *Amblyomma*

Amblyomma americanum: Lone star tick
Amblyomma cajennense: Cayenne tick
Amblyomma dissimile: Iguana tick
Amblyomma hebraeum: Tropical African Bont tick
Amblyomma maculatum: Gulf Coast tick
Amblyomma tuberculatum: Gopher Tortoise tick
Amblyomma variegatum: Variegated tick/Tropical Bont tick

Genus *Argas* (Soft multihost tick)

Argas persicus: Fowl tampan/fowl tick/poultry tick

Genus *Dermacentor*

Dermacentor albipictus: Winter tick/Moose tick
Dermacentor andersoni: Rocky Mountain Wood tick

Dermacentor nigrolineatus: Brown winter tick
Dermacentor nitens: Tropical horse tick
Dermacentor occidentalis: Pacific Coast tick
Dermacentor reticulatus: Marsh tick
Dermacentor variabilis: American dog tick

Genus *Haemaphysalis*

Haemaphysalis chordeilis: Bird tick
Haemaphysalis leachi: Yellow dog tick
Haemaphysalis leporispalustris: Rabbit tick
Haemaphysalis longicornis: New Zealand cattle tick or Bush tick

Genus *Hyalomma*

Hyalomma detritum: Bont legged tick
Hyalomma dromedarii: Camel tick
Hyalomma excavatum: Brown ear tick
Hyalomma marginatum: Mediterranean tick
Hyalomma truncatum: Bont legged tick

Genus *Ixodes*

Ixodes angustus: Round tick
Ixodes canisuga: British dog tick
Ixodes hexagonus: Hedgehog tick
Ixodes holocyclus: Paralysis tick
Ixodes pacificus: California blacklegged tick
Ixodes persulcatus: "Taiga" tick
Ixodes pilosus: Sour veld/Russet/Sheep paralysis/Bush tick
Ixodes ricinus: Castor bean tick/Sheep tick
Ixodes rubicundus: Paralysis tick
Ixodes scapularis: Shoulder tick or Blacklegged tick, Deer tick

Genus *Margaropus*

Margaropus reidi: Sudanese beady-legged tick

Genus *Ornithodoros*

Ornithodoros moubata: The sand eyeless tampan
Ornithodoros savignyi: Sand tampan or African eyed tampan

Genus *Otobius*

Otobius megnini: Spinose ear tick

Genus *Rhipicephalus*

Rhipicephalus (Boophilus) annulatus: North American tick/Cattle fever tick
Rhipicephalus appendiculatus: Brown ear tick
Rhipicephalus capensis: Cape brown tick
Rhipicephalus (Boophilus) decoloratus: Blue tick
Rhipicephalus evertsi: Red-legged tick
Rhipicephalus (Boophilus) microplus: Tropical fever tick/Cattle tick
Rhipicephalus pravus: Brown tick
Rhipicephalus pulchillus: Zebra tick
Rhipicephalus sanguineus: Kennel tick/Brown dog tick/Tropical dog tick
Rhipicephalus simus: The glossy brown tick

GENERAL MORPHOLOGY OF TICKS

Though typically similar with other organisms, the peculiarities in morphological characters of ticks clearly reveal their adaptation as hematophagous ectoparasites. Ticks, to some extent, are morphologically similar to other acari but they can be clearly distinguished by certain characters, which include:

1. A hypostome with backward pointing teeth for securing to its host (Plate 1.1).
2. Haller's organ, a complex sensory organ with setae, situated on the dorsal side of tarsus I (Plate 1.2).
3. A pair of stigmata located posterior to coxa IV or dorsal to coxa III and IV (Plate 1.2).
4. Palps with three or four segments (Plate 1.1).
5. Digits of chelicerae with dentate faces directed externally (Plate 1.1).

External Structures

The Integument

The integument of argasids is tough and leathery whereas the ixodids possess a hard chitinous exoskeleton.

Body Segmentation

The body of the tick is divided into two portions—an anterior gnathosoma and a posterior idiosoma. The gnathosoma consists of a basis capituli bearing a pair of pedipalps, a pair of chelicerae and a hypostome

Plate 1.1 Ventral view of the capitulum of a hard tick showing the chelicera, hypostome, and palps.

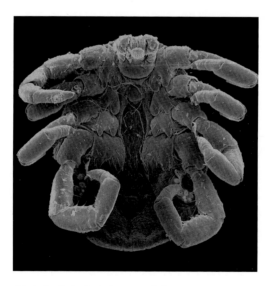

Plate 1.2 Ventral view of the body of a hard tick showing Haller's organ, spiracular plate, anal orifice and groove, postanal median groove, coxa I–IV and coxal spurs.

(Plate 1.1). The body posterior to the gnathosoma is called the idiosoma, which bears four pairs of legs.

The body segmentation of acarids in comparison to other arachnids is shown in Table 1.4.

Table 1.4 Body Segmentation of Acarids Compared to Other Arachnids

Tagmata	Structures Present	Subclass Acari	Other Arachnids
Gnathosoma (head)	*Chelicerae* (modified first appendages) *Pedipalps* (modified second pair of appendages)	*Gnathosoma* (head)	*Prosoma* (cephalothorax) (gnathosoma + podosoma)
Podosoma (thorax)	Four pairs of walking legs	*Idiosoma* (podosoma + opisthosoma)	
Opisthosoma (abdomen)	Respiratory organs		*Opisthosoma* (abdomen)
	No appendages		
Diagrammatic representation of body segments			

The Capitulum (Head or Gnathosoma)

The anterior most tagma or gnathosoma of ticks comprising the mouthparts is known as the capitulum. In hard ticks, it extends forward bearing resemblance to a true head. In soft ticks, the capitulum is situated on the ventral side in a depression known as the camerostome and not visible from the dorsal aspect.

The capitulum is made of the following parts:

Basis capituli: This is the basal part of the capitulum (Fig. 1.2). It may be hexagonal (when the lateral margins are angular) or rectangular (when the lateral margins are straight). The dorsal and ventral surfaces possess dorsal and ventral ridges respectively. The basis capituli may have distinct or indistinct lateral bulges on the ventral surface called auricles and horn-like postero-lateral projections or cornua. On the dorsal surface of the basis capituli are the porose areas (areae porosae). These are present only in female hard ticks.

Chelicerae: Anterior to the basis capituli is a pair of chitinous cylindrical shaft-like structures called chelicerae (Plate 1.1). These structures are enclosed in a single sheath. Each shaft is armed with laterally directed denticles.

Hypostome: This single median spatulate structure is an anterior projection from the basis capituli. The hypostome is situated in

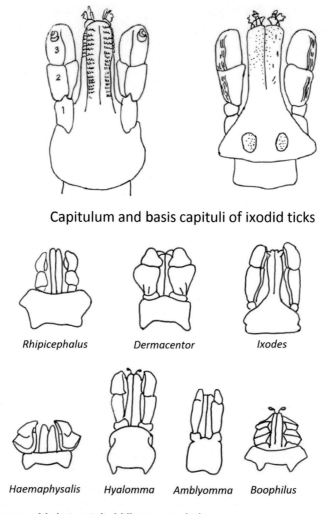

Capitulum and basis capituli of ixodid ticks

Rhipicephalus   Dermacentor   Ixodes

Haemaphysalis   Hyalomma   Amblyomma   Boophilus

Figure 1.2 Structures of the basis capituli of different genera of ticks.

close association and ventral to the cheliceral sheath. Rows of recurved teeth on the ventral surface of hypostome are used to anchor the tick to its host.

Pedipalps: Also known as palps, is a pair of finger-like projections located laterally on either side of hypostome (Plate 1.1). Each pedipalp comprises four articles (I–IV). Characteristic structures of the internal margin of Article I are useful in identification of different ixodid species. Article IV is very small and possesses a sensory organ. The palps may have projections and saliences in addition to bristles, hairs and conspicuous spurs.

The Idiosoma (or Body)

The idiosoma or body is the large region behind the gnathosoma (head) and bears the legs, various orifices, sensory and other structures. These are

Scutum or conscutum: Sexual dimorphism of adult hard ticks can be readily distinguished by the presence of a scutum. This is a chitinous hard plate covering almost the entire dorsal surface of male ixodid ticks (conscutum). This plate covers only the anterior third of the dorsal surface of unfed female hard ticks (larvae, nymphs, and adults) and is called the *scutum* (Plate 1.3). The characteristic shape, color, and setation of the scutum help to identify the genus of ticks. The scutum and conscutum are absent in argasid ticks. In some species of *Amblyomma*, the scutum is ornate, i.e., attractively ornamented with various designs of glittering color. The scutum is also the site of attachment of various muscles that control the chelicera and other parts of the tick's body.

Alloscutum: Posterior to the scutum is the alloscutum (Plate 1.3). This structure is highly extensible and allows for accommodation of blood by expanding when the female larva, nymph and adult tick feeds.

Pseudoscutum: This is a large elevated area present at the anterior part of the scutum in some ticks.

Scapulae: scapulae or shoulder blades are antero-lateral angles projected on either side of the scutum or conscutum.

Plate 1.3 Dorsal view of a female hard tick showing the scutum, alloscutum, cervical and lateral grooves.

Eyes: Some tick species possess eyes. These are two small protuberances situated on the antero-lateral margins of the scutum. Their shapes differ with species. They may be small or large, flat or convex. The ticks of the genera *Ixodes* and *Haemaphysalis* have no eyes, whereas *Rhipicephalus* and *Hyalomma* species possess indistinct or convex eyes. In argasid ticks, the eyes are present only in some species of the genus *Ornithodoros* and are located on the lateral body surface.

The porose areas: These are two small circular/oval depressions or pits on the dorsal surface of the basis capituli posterior to the palps of female hard ticks (Plate 1.4). These areas contain numerous pores, which secrete a waxy substance involved in protection of newly laid eggs.

Spots and stripes: These are patches of pigments of various sizes present on the scutum of ornate ticks.

Punctations: These are small and circular shallow pits distributed all over the scutum or conscutum. These may be sparse, dense, or localized according to the genera of ticks.

Festoons: These are a chain of small chitinous plates on the posterior margin of the dorsum of some genera of hard ticks. These are pale or dark in color, with/without enamel patches. They are usually 9−11 in number and are not distinct in engorged females (Plate 1.5).

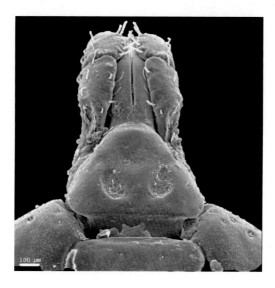

Plate 1.4 Dorsal view of the basis capituli of a hard tick showing the porose areas.

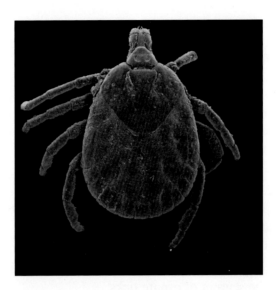

Plate 1.5 Dorsal view of a female hard tick showing the festoons.

Caudal process: This is the pointed posterior margin of the body present in males of some species of ticks. The caudal process may be short or long, narrow or broad.

Genital orifice or aperture: This structure is situated ventrally on the median line posterior to the basis capituli. The position of the genital aperture can vary with species. It is generally located between coxae III or IV according to the genus (Plate 1.6). The aperture is bounded anteriorly by a chitinous flap or postgenital plate and posteriorly by a ridge-like elevated area of the integument. In the soft tick (family: Argasidae) the genital opening is slit-like in females and half-moon shaped in males. Voiding of eggs in females occur through this aperture.

The anus:Lies on the median line of the ventral surface of the body, posterior to the genital orifice. The anal orifice is enclosed by a small ring or annulus guarded by a valve on each side (Plate 1.6).

The stigmata or spiracles: In soft ticks, the spiracles lie between coxae III and IV. In the hard ticks, they lie posterior to coxa IV (Plate 1.2). Spiracles are surrounded by spiracular plates and are the external openings of the tick's respiratory system.

Coxal pores: Coxal pores are found in nymph and adult soft ticks. These are the openings of the coxal glands and located between coxae I and II. Excess fluids filtered from the blood meal are excreted through these pores.

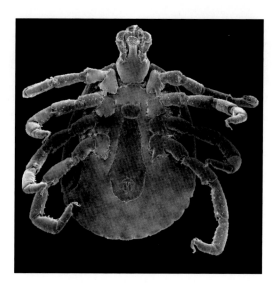

Plate 1.6 Ventral view of a female hard tick showing the legs, anal groove, anus, and genital orifice.

Plates or shields: In certain species of male ticks, a number of chitinous shields or plates are present symmetrically and posterior to the anus. The smaller structures are plates and larger ones are shields.

The legs: Tick larvae have six legs whereas nymphs and adults have eight. Each leg is composed of six segments: coxa, trochanter, femur, tibia, pro-tarsus, and tarsus (Plate 1.6). Some of the segments may have annulations or pseudo-articulations. Spurs are conspicuous processes on the coxae, which may also present on the tarsus. Attached to each tarsus is a pair of claws, pad-like pulvilli and an empodium. Pulvilli are adhesive in nature, which enables the tick to climb on many surfaces.

The outer border of the tarsus on the first pair of legs has a small, pitted structure called Haller's organ. This is an olfactory organ composed of an array of sensory setae and is sensitive to humidity and aids in finding a host and feeding (Plate 1.2).

Grooves: Grooves are present both on the dorsal and ventral sides. On the dorsum, these are long channels. The position and size of the grooves facilitate identification of ticks. Those of importance for identification purposes are the cervical and lateral grooves (Plate 1.3). On the ventral side two genital grooves extend posteriorly from the genital aperture to near the anus. The anal groove is a single groove that contours the anus anteriorly. In some species, the anal groove is absent but when present it is valued for taxonomic importance (Plate 1.6).

Internal Structures

The Alimentary Canal

This is a straight tube. The mouth is formed by the apposition of the chelicerae and hypostome. The pharynx and esophagus are small and narrow. The wide mid-intestine has caeca appendages at its two ends. These are distended with blood after feeding and serve as a storehouse in times of little or no food. The mid-intestine of some ticks contains anticoagulin (an anticoagulant) which prevents or delays the clotting of the host's blood when the tick feeds.

The hindgut is small and very narrow and terminates in the rectal sacs. Two very long malpighian tubules open into the rectum. The salivary glands are large, elongated and grape-like bodies formed on each side. The common salivary duct opens in the mouth near the base of hypostome. The tick's salivary secretions are strongly hemolytic.

Respiratory Organs

The respiratory organs are the tracheal trunks (usually five) opening to the outside through a pair of spiracles. Internally the tracheal trunks further divide into smaller branches or tracheoles. The spiracles are small plates situated ventrolaterally (one on each side) on the body of the tick. Hinton[37] studied the structure of the spiracles of *R. (B.) microplus*. Schol et al.[38] studied the morphology of spiracles in adult *H. truncatum*. In these ticks, the spiracular plate forms the outer part of the spiracle. In case of the family Ixodidae, the spiracle is situated posterior to coxa IV and in Argasidae between coxae III and IV (Plate 1.2).

Taxonomists use the shape of spiracular plates for identification of tick species.

Heart, Arterial Vessels, Sinuses, and Hemocoel

Ticks have an open circulatory system, which consists of the heart, arterial vessels, sinuses, and the hemocoel. The sinuses surround the central nerve trunk, esophagus and nerves to the mouthparts anteriorly and pedal nerve trunks laterally.

The hemolymph flows from the heart to the anterior dorsal aorta, then to the perineural sinus, which encloses the central nerve trunk. The hemolymph then flows to the sinuses enclosing the major nerve trunks and back to the hemocoel from the lacunae within appendages and then to the heart through two pairs of ostia.[39,40]

Reproductive Organs

Female

The female reproductive organs of ixodid and argasid ticks are similar[41] consisting of an ovary, paired oviducts, a uterus, vagina, tubular accessory glands and Gene's organ. But the porose areas are absent in argasids.

Ovary: The ovary is an elongated single organ that lies above the posterior intestinal diverticulum.

Oviducts: These are paired organs, which then terminate into a common oviduct or uterus.

Uterus: This organ is sac-like and lies anterior to the ovary in the middle of the body of the tick.

Vagina: The uterus continues through a short tube, the vagina, into the external genital opening.

The accessory glands: These are two sausage-shaped glands opening into the vagina. In some species of ticks, these glands are absent.

Gene's organ: This is an eversible sac-like organ with finger-like lobes. It emerges through an aperture situated immediately anterior to the dorsal aspect of basis capituli in camerostomal folds of ixodid ticks or in a similar location in argasids. Gene's organ excretes a lipid rich, waterproof layer, which coats the eggs during oviposition. This layer prevents desiccation of the tick embryo before hatching.

The porose areas: Two porose areas with groups of pores are present on the tectum of the basis capituli of all female ixodids except *Ixodes kopsteini*.[42] It is thought that the secretions from these areas inhibit the autoxidation of the egg-wax secreted by Gene's organ.

Male

The reproductive organs of male ticks consist of testes, vas deferens, seminal vesicles, ejaculatory duct, accessory glands, and coxal glands.

The testes: The paired testes lie across the posterior part of the body. They are thin-walled tubes with an irregular outline.

The vas deferens: A continuation of the testes.

Seminal vesicle: A large sac-like organ that receives the end of the vas deferens.

Ejaculatory duct: The continuation of the seminal vesicle and continues into the genital opening.

Accessory glands: Also known as "white glands" due to their chalky color, constitute a large mass of a variable number of lobes.

Coxal glands: These paired organs are excretory in nature each located in close proximity to coxae II. The opening duct of the gland is situated on the ventro-posterior part of coxa I.

The Ganglia and Nerve Trunks

The ganglia of the central nervous system of ticks are highly condensed into a mass called the peri-esophageal ganglion or synganglion, which is enclosed, in a circulatory sinus. The synganglion is divided into two by the esophagus. From the pre-esophageal portion, arise paired optic, cheliceral, and pedipalpal nerves and an unpaired stomodeal or pharyngeal nerve. From the post-esophageal part arise four pairs of pedal nerve trunks serving the four pairs of legs and four major pairs of opisthosomal nerves. Fine "sympathetic" nerves connect all four pedal nerve trunks laterally on each side of the synganglion.[40]

HABITS AND GENERAL LIFE HISTORY OF TICKS

Life Cycle of Hard Ticks

The life cycle of hard ticks consist of four stages (egg, larva, nymph, and adult) compared to more stages in soft ticks (Fig. 1.3). Eggs of hard ticks are laid in large numbers in cracks and crevices or in leaf mats under bushes or grass adjacent to where the female drops off the host. A single female ixodid tick may deposit as many as 15,000 eggs. The eggs are proportionally large, round, and brown when ripe. Hard ticks deposit all their eggs in a single act of oviposition after which the female dies. Hexapod larvae emerge from eggs and feed for a week or more after hatching. The larvae of most species climb onto grass or small shrubs wherein they form a cluster and await a susceptible host in an act known as questing. Larvae may transfer onto any passing animal/human but only attach to feed upon the preferred host. Ticks may feed on unsuitable hosts in the absence of a preferred one. On the suitable host, the larva immediately attaches itself and takes a blood meal for several days. When engorged, it drops to the ground, seeks shelter and remains quiescent. During this period, digestion of the engorged blood occurs and the larva undergoes metamorphosis to form a nymph. Nymphs do not form clusters but attach singly to a new host, feed and drop off on the ground where they molt into an adult. Mating may take place either on or off the host. After

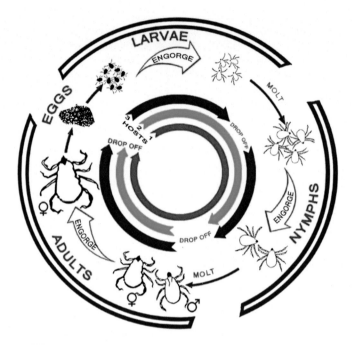

Figure 1.3 General life cycle of one-, two-, and three-host hard ticks.

fertilization, the adult female tick attaches to the host and feeds until fully engorged. One adult ixodid female can take up to 2 mL of blood. After a blood meal, the engorged female may weigh up to 250 times her unfed weight.[43] In many instances, copulation will have a marked effect on the subsequent engorgement of the female. The female then drops off on the ground, takes shelter, and starts laying eggs.

Ixodid ticks are characterized as one-, two-, or three-host ticks based on their feeding habits (Table 1.5).

Life cycle of one-host ticks: The larva, nymph, and adult feed only on one host. The larva remains attached, feeds on the same animal, molts into a nymph then an adult. The adult continues to feed and drops off when fully engorged, e.g., all *Rhipicephalus (Boophilus)* spp. (five species), *Margaropus* spp. (three species), and *Dermacentor* spp. (two species) are one-host ticks.

Life cycle of two-host ticks: The larva molts into a nymph and both stages feed on the same host. After feeding, the nymph drops off from the host and molts to an adult on the ground. The newly

Hosts	One-Host Tick	Two-Host Tick	Three-Host Tick
Table 1.5 The Number of Hosts Required for Feeding of One-, Two- and Three-Host Ticks at Different Stages			
Host 1	Larva Nymph Adult	Larva Nymph	Larva
Host 2	–	Adult	Nymph
Host 3	–	–	Adult

emerged adult finds a second host on which to feed, e.g., a few *Hyalomma* and *Rhipicephalus* species.

Life cycle of three-host ticks: These ticks require three animals to complete their life cycle. For each stage, the larva, the nymph, and adult drop off the host after feeding to molt or oviposit on the ground. Of 650 species of hard ticks, 600 species have three-host life cycles, e.g., *A. hebraeum, R. appendiculatus*.

The Stages of Life Cycle

Egg: After engorgement, the female tick drops from the host and takes shelter in cracks and crevices, where it lays its eggs. The eggs are in clusters and appear light orange in color under the dissecting microscope, which gradually turn grayish in color when they hatch. Eggs that are darker in color do not hatch. The eggshells of most ticks are two-layered on one side, and three-layered on the other.[44] When rearing ticks in the laboratory, removal of the dead female ticks after oviposition is necessary to prevent contamination of the laid eggs with pathogenic organisms from the body. Contamination prevents development of seed ticks from eggs. The six-legged larvae or seed ticks hatch from the eggs through a longitudinal slit (Plate 1.7) within 20–30 days of oviposition depending upon the humidity and temperature.

Seed tick/Larva: The larvae have a tendency to remain in a clump without much activity. If disturbed, they resume their formation. They become active after 3–4 days, and move around in search of a suitable host. In a tick-infested pasture, larval ticks congregate on the lower surface of the tip of grass blades. It has been observed that the larval ticks do not attach easily when put on the host immediately after hatching. However, they attach in no time if put on the host after 3–4 days of hatching, when they are hungry. They may select a

Plate 1.7 Longitudinal slit of egg shell of tick

place for attachment on the host's body and change position after a short while to find another suitable site. At the time of attachment of the larval ticks, initially the host gets irritated and tries to scratch, but after about 30–45 minutes of attachment, no sign of irritation is observed. The seed tick feeds on host blood for several days. After a blood meal, the larvae of one- and two-host ticks molt on the host's body, whereas those of three-host ticks drop off on the ground and molt there. In all instances, the exoskeleton of the larvae is shed and a slightly larger eight-legged nymph emerges.

Nymph: In case of one- and two-host ticks, the larvae and the nymphs continue to remain and feed on the same host. One-host ticks remain on the same host after molting to the adult stage. In two-host ticks, after sucking blood, the nymph drops off from the host, molts into an adult on the ground and then climbs onto a second host. In case of three-host ticks, the nymph molts from the dropped off larva, climbs onto a second host, sucks blood, and drops off again to transform into an adult which attaches to a third host.

Adult: After emergence from the nymph, the adult starts feeding on the host's blood. Fertilization occurs during the early days of feeding. After fertilization, the female hard tick continues to feed to reach several to 100 times its unfed body weight depending upon the tick species.[45] The male remains attached to the host for a longer period than females (several days to months) and suck small amounts of blood meal compared to females. The females emit a pheromone that attracts the male, which can mate with several females while on the host. It is interesting to note that during the period of engorgement maximum blood is sucked in the last 1 or 2 days before leaving the host. The engorged female then moves to a suitable, undisturbed location (cracks and crevices especially around animal dwellings) where it begins laying eggs on the second or third day. A previous study done on *R. (B.)microplus*, revealed that the number of eggs laid per day gradually increased until the 9th to 12th day and then decreased gradually.[46] The duration of oviposition is about 20−30 days in the rainy season, which is the ideal time for multiplication of ticks. Both temperature and humidity are important factors in tick multiplication. It is evident that the production of eggs increases both with suitable temperature and humidity.[46] After completion of egg laying, the female tick dies within 2−4 days.

Pattern of egg laying

Ticks have host preferences. Cattle ticks prefer cattle blood. Experiments have revealed that the average weight of female cattle ticks engorged from cattle was higher than the weight of those engorged from rabbits.[46] Likewise, cattle ticks, fed on cattle, oviposited more eggs over a longer period than those fed on rabbits. The total number of eggs laid by the female is directly proportional to the weight of the engorged ticks. Those weighing less than 70 mg are not able to produce eggs or produce very few eggs that do not hatch.[46]

Life Cycle of Soft Ticks

Compared to hard ticks, soft ticks have more stages (egg, larva, 2−5 nymphal stages, and adult) in their life cycle (Fig. 1.4). Most soft ticks are multi-host ticks. Females feed rapidly and drop off the host after engorgement. Fully engorged female ticks lay eggs on the ground especially in cracks and crevices. Each female tick produces a small batch of eggs (100−500).[36] This phenomenon of feeding and egg laying is repeated for up to six times. Each egg is attached to the other by a narrow bond (Plate 1.8). Initially the eggs are light orange in color and

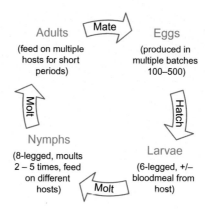

Figure 1.4 General life cycle of soft ticks.

Plate 1.8 Soft tick eggs attached to each other.

gradually turn gray just before hatching (Plate 1.9). Darker colored eggs do not hatch. Six-legged larvae hatch from the eggs within 17–23 days in the case of *O. savignyi*[47] through a longitudinal opening. The period of hatching from the egg varies with species, humidity, and temperature. The larva attaches to the host, sucks blood, drops to the ground and transforms to an eight-legged nymph. The larvae of *O. savignyi* do not need a blood meal to transform to nymphs. Several nymphal stages (usually 2–5) occur in argasids. Each nymphal stage feeds on a blood meal, drops off the host, molts to a further nymphal stage and again climbs onto a host. Similar feedings follow on different

Plate 1.9 Freshly laid tick eggs.

Plate 1.10 Soft ticks attached to the fetlock region of a cow.

individual hosts and finally the nymph molts into an adult on the ground. Both adult males and females show multiple feeding. In Nigeria, it has been observed that *O. savignyi* seldom climbed beyond the height of the fetlock region of cattle (Plate 1.10) and attacked humans (Plate 1.11).

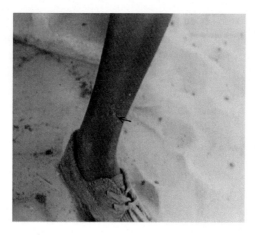

Plate 1.11 Soft ticks attached to the lower leg of a human.

FEEDING MECHANISM OF TICKS

As hematophagous ectoparasites, all stages of ticks feed only on blood of their hosts. The tick climbs onto the host and crawls to the preferred site of attachment. It holds the skin with the help of its tarsal claws and then attach to the skin with the mouthparts. Only the hypostome penetrates the skin. The sharp cutting plates on the chelicerae help to cut a hole in the dermis and break the capillary blood vessel very close to the surface of the skin, forming a feeding lesion. The hypostome is inserted through the opening made. The palps remain outside the wound and spread horizontally on the skin surface. The hypostomal teeth help the tick to anchor itself to the host during feeding and the tick begins drawing blood up to its mouth through a grooved channel that lies in between the chelicerae and hypostome. A cement-like substance (cementum) is secreted into the wound in the first 5−30 minutes by some ixodid ticks when they feed. This material hardens quickly around the inserted mouthparts, which also helps to stabilize the tick to its host when feeding.[11]

SEASONAL ACTIVITY OF TICKS

Ticks and tick-borne diseases (TBDs) are widely distributed throughout the world, particularly in tropical and subtropical regions. The incidence of tick species and TBDs varies with the ecology of the regions. The dynamics of transmission of tick-borne pathogens is also influenced by seasonal changes.[48] Seasonal activity of ticks varies with species and country.

The effect of season on the incidence of ticks is a very important aspect of their biology. Accurate knowledge of the combined effect of variable temperature, humidity, and rainfall on the tick population are essential for determining the multiplication of ticks on and off the host, which in turn is vital for planning control measures through management and application of ixodicides at the correct time.

Tick numbers are highest during the rainy season followed by summer and winter. Both high temperature ($\geq 28\,°C$) and relative humidity ($\geq 75\%$) are essential for the multiplication of ticks. The summer, with high temperature and low humidity, is not suitable for tick multiplication. Conversely, low temperature and humidity in winter brings down the population to the lowest level.

Newson[49] observed that *R. appendiculatus* populations increased with significant rainfall in a number of ecologically distinct sites. Control measures, especially application of ixodicides should be adopted during the rainy season to destroy the maximum number of ticks and to stop their multiplication.

Trinidad and Tobago have two main seasons. The dry season is from January to May and the wet or rainy season, from June to December. The minimum and maximum temperatures, relative humidity, and rainfall indicate that the overall atmosphere is very conducive for the multiplication of ticks year-round on these two islands (Figs. 1.5–1.7).

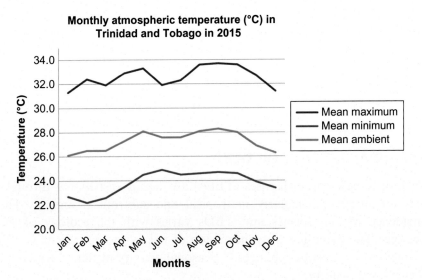

Figure 1.5 Graph of monthly atmospheric temperature for 2015 in Trinidad and Tobago.

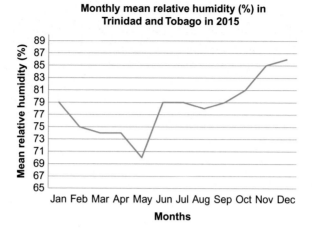

Figure 1.6 Graph of monthly relative humidity for 2015 in Trinidad and Tobago.

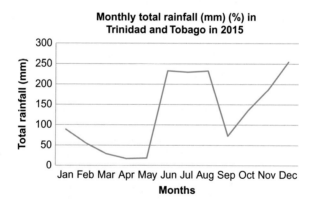

Figure 1.7 Graph of monthly total rainfall for 2015 in Trinidad and Tobago.

METHODS OF TICK CONTROL

These include off-host and on-host treatment. Off-host control methods include burning pastures, cultivation of lands, starvation, use of grasses as repellents, use of natural enemies, and sterile hybrid techniques.[50] On-host control measures include the use of synthetic acaricides and phytochemicals as well as immunological control methods. The chemicals are applied by dipping, spraying, swabbing, neckbands, ear-tag/collars, pour-on and baths. The method of handpicking of ticks is suitable when the animals and ticks are few in number.

Chemical Methods of Tick Control

Application of acaricides is adopted during the rainy season to destroy the maximum number of ticks and to stop their multiplication. The strategic treatment schedule is applied, once at the beginning of the rainy season to stop their multiplication and again at the end of the rainy season to eliminate those ticks which survived the first treatment and multiplied during the rainy season.

Tick control with chemical acaricides was very popular at one time and was partially successful, but the drawbacks of using synthetic acaricides include harmful residual effects on meat and milk for human and animal consumption. Continuous use of acaricides also results in the development of resistant tick strains. Ticks resistant to the most commonly used acaricides including macrocyclic lactones,[51] synthetic pyrethroids,[52–55] benzene hexachloride (BHC),[56] carbamates,[57] and organophosphates[58–60] have been reported. Recently Abbas[61] has reviewed the development of acaricidal resistance in cattle ticks. Patterns of resistance to acaricides in some populations of *R. (B.) microplus* ticks from Jamaica, West Indies were reported.[62]

Pharmaceutical companies are not encouraged to bring newer acaricides to the market because their development is a long and expensive process, and resistance develops among the ticks (especially one-host ticks) against the new drugs within a few years due to continuous application.

Biological Methods of Tick Control

Biological control of ticks appears to be feasible. They include the introduction of natural predators (e.g., ants, beetles, and spiders), parasites (e.g., mites, insects, and nematodes), and bacterial and fungal pathogens. Other methods include releasing male ticks sterilized by irradiation or hybridization and the immunization of hosts against ticks. Integrated pest management can be used to control ticks. In this system, a number of different control methods are adapted against a specific tick species or infested location with consideration for any effects on the environment.

Immune Response

The development of drug resistance among ticks and the harmful effects of acaricides on animals, humans and the environment have encouraged researchers to study the immune response and immune-prophylaxis for tick infestation. Immune response to tick infestation has been studied by various researchers.[63–74]

Immune response in the host may follow either after natural infection or after introduction of the agent, live or inactivated. The response in the host may be cell-mediated or humoral. Such immunity may be demonstrated by in-vitro or in-vivo testing. In the case of parasitic immunity, the immune response can be observed in the host and in the parasite when introduced into the immune host by observing its reduced ability to thrive or reproduce. There are several methods for qualitative and quantitative estimation of humoral and cell-mediated immune responses. Of all the tests, demonstration of protective immunity is the most important, as the ultimate object of immunoprophylaxis is to achieve protective immunity.

Recent studies have shown that vaccination of hosts with selected tick proteins can reduce tick feeding, their reproduction, and even the infection and transmission of TBDs.[75]

Vaccines Against Ticks
Since the last quarter of the 20th century, identification of protective antigens against ticks and research on immunological approaches to the control of ticks via the use of vaccines has been employed worldwide.[76]

Tick vaccines such as TickGARD and Gavac, containing the midgut membrane-bound recombinant protein BM86 of *R. (B.) microplus* have been used in Australia and in Latin American countries (Cuba, Colombia, Dominican Republic, Brazil, and Mexico) for the control of ticks. Vaccination of cattle with BM86 was found to reduce the number, weight, and fecundity of engorged female ticks.[77,78] As a result, the incidence of babesiosis and tick infestations of vaccinated cattle herds decreased. In spite of positive results of this BM86 vaccine, strain-to-strain variation in efficacy exists and *Rhipicephalus* species seem to be affected most. The mechanism by which BM86 functions is not fully understood.

In Cuba, the cost effectiveness analysis showed a 60% reduction in the number of acaricide treatments, together with the control of tick infestation and transmission of babesiosis.[78] In Cuba, a previous report on the "Effect of application of Gavac TM plus integrated tick control on acaricide treatments" on 668,000 animals revealed reduction in frequency of dips from 14−21 days to 60−180 days (>1000 days at certain locations), reduction in chemical consumption from 371 tons annual average to 154 tons in 2009, and reduction of incidence of hemoparasites of 98%.[79] It was concluded that vaccination is the best method that is environmentally friendly for the control of tick infestations and cheaper than other control methods.

COLLECTION AND PRESERVATION

Collection

Ixodid ticks live on the bodies of domestic and wild animals and on vegetation. Therefore, collection of ticks from these sources (animals and vegetation) is commonly used. Two methods used for collecting questing nymphs and adults from the environment are flagging and dragging. A drag consists of a square piece of light-colored, rough textured cloth of about 1 m^2 attached by one side to a pole. The method involves dragging the cloth behind the investigator, by a rope attached to both ends of the pole. The drag is kept horizontally to facilitate contact of ticks to the cloth. The drag cloth is inspected frequently for ticks and the attached ticks are removed and preserved. This dragging method is suitable for collection of ticks from low vegetation.

Flagging is similar to dragging but involves a smaller cloth, attached like a flag to one end of a pole with the other end used as a handle. The flag is swept through the vegetation with thick brush and shrubs.

Ticks are collected directly from animals by handpicking. This method consists of manually searching for ticks on the bodies of domestic and wild animals. Adults and immature stages are removed by hand using forceps and preserved.

Preservation

Ticks are routinely preserved in glass vials containing 70% ethanol, which prevents hardening of the tegument. Addition of a few drops of glycerin aids in maintaining the natural color of the ticks. Occasional replacement of 70% ethanol to vials containing ticks is advised to avoid reduction of the alcohol concentration.

PROCEDURE FOR EGG COUNTING

Ticks are kept in clean petri dishes during egg laying. A clean needle is used to place eggs on a slide with a square grid for counting. Eggs are examined and counted using a dissecting microscope (Plate 1.12).

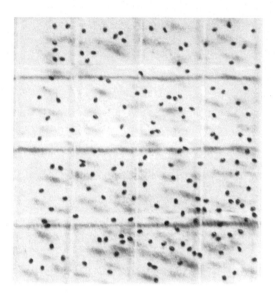

Plate 1.12 Tick eggs on the counting grid.

CHAPTER 2

Ticks in the Caribbean Region

A list of tick species present in Trinidad and Tobago and their role in the transmission of tick-borne diseases have been compiled.[80] Cruz[81] recorded 45 tick species from 11 genera and 2 families from the West Indies. This comprises 14 species of *Ornithodoros*, 10 species of *Antricola*, 9 species of *Amblyomma*, 3 species each of *Argas* and *Rhipicephalus*, 2 species of *Ixodes*, and 1 species each of *Parantricola*, *Dermacentor* (*Anocentor*), *Haemaphysalis*, and *Aponomma*. Ticks reported from the Caribbean countries have been listed in Table 2.1.[80–87] Based on a study of the vector situation of tick-borne diseases, Camus and Barré[85] mentioned that the most important ticks

Table 2.1 Ticks recorded from the Caribbean countries, West Indies[80–87]	
Tick Species	Countries Reported From
Argas miniatus	Cuba, Puerto Rico, Jamaica, Antigua and Barbuda, Martinique, Trinidad and Tobago
Argas persicus	Puerto Rico, Cuba, Antigua and Barbuda, Barbados, Trinidad and Tobago, Martinique, Jamaica
Argas radiatus	Cuba
Antricola armasi	Cuba
Antricola centralis	Cuba
Antricola cerny	Cuba
Antricola granai	Cuba
Antricola habanensis	Cuba
Antricola martelorum	Cuba
Antricola naomiae	Cuba
Antricola occidentalis	Cuba
Antricola siboney	Cuba
Antricola silvai	Cuba
Parantricola marginatus	Cuba, Puerto Rico

(Continued)

Ticks of Trinidad and Tobago—An Overview. DOI: http://dx.doi.org/10.1016/B978-0-12-809744-1.00002-5

Table 2.1 (Continued)

Tick Species	Countries Reported From
Ornithodoros azteci	Trinidad and Tobago, Jamaica, Cuba
Ornithodoros brody	Cuba
Ornithodoros capensis	Trinidad and Tobago, Saint Martin, Dominica, Jamaica, Cuba
Ornithodoros cyclurae	Cuba
Ornithodoros denmarki	Trinidad and Tobago, Martinique, Dominica, Guadeloupe, Jamaica, Cuba
Ornithodoros dusbabeki	Cuba
Ornithodoros elongatus	Dominican Republic
Ornithodoros hasei	Martinique, Guadeloupe, Barbuda, Trinidad and Tobago, Dominica
Ornithodoros kelleyi	Cuba
Ornithodoros natalinus	Cuba
Ornithodoros puertoricensis	Jamaica, Guadeloupe, Virgin Islands, Puerto Rico, Trinidad and Tobago
Ornithodoros tadaridae	Cuba
Ornithodoros talaje	Puerto Rico
Ornithodoros turicata	Jamaica
Ornithodoros viguerasi	Trinidad and Tobago, Jamaica, Puerto Rico, Cuba
Amblyomma albopictum	Cuba, Cayman Islands, Dominica
Amblyomma antillorum	Virgin islands, Dominica, Bahamas, Turks and Caicos Islands, British Virgin Islands
Amblyomma arianae	Puerto Rico, Cayman Islands
Amblyomma auricularium	Trinidad and Tobago
Amblyomma cajennense	Cuba, Jamaica, Trinidad and Tobago,
Amblyomma calcaratum	Trinidad and Tobago
Amblyomma cruciferum	Puerto Rico
Amblyomma dissimile	Cuba, Jamaica, Puerto Rico, Barbados, Grenada, St, Lucia, Antigua, Trinidad and Tobago, Antigua and Barbuda
Amblyomma goeldii	Jamaica
Amblyomma hebraeum	Antigua/Barbuda
Amblyomma humerale	Trinidad and Tobago
Amblyomma longirostre	Jamaica, Trinidad and Tobago
Amblyomma nodosum	Trinidad and Tobago
Amblyomma ovale	Trinidad and Tobago
Amblyomma rotundatum	Montserrat, Jamaica, Grenada, Guadeloupe, Martinique, Trinidad and Tobago
Amblyomma torrei	Cuba, Puerto Rico, Cayman Island
Amblyomma variegatum	Anguilla, Puerto Rico, Antigua/Barbuda, Barbados, Dominica, Martinique, Guadeloupe, St. Kitts, Nevis, Montserrat, St. Lucia, St. Vincent, British Virgin Islands, Saint Marten
Anocentor nitens	Cuba, Jamaica, Puerto Rico, Virgin Islands, St. Kitts, St. Vincent, Montserrat, Dominica, St. Martin, Antigua /Barbuda, Guadeloupe, Dominica, Bahamas, Belize, Martinique, Trinidad and Tobago, Haiti, Grenada, Barbados

(Continued)

Table 2.1 (Continued)	
Tick Species	Countries Reported From
Aponomma quadricavum	Cuba
Haemaphysalis juxtakochi	Trinidad and Tobago
Haemaphysalis leporispalustris	Cuba
Ixodes capromydis	Cuba
Ixodes downsi	Trinidad and Tobago
Ixodes luciae	Trinidad and Tobago
Rhipicephalus (Boophilus) annulatus	Antigua/Barbuda, Bahamas, St. Kitts, Jamaica, Puerto Rico, Guadeloupe
Rhipicephalus (Boophilus) microplus	Widely distribution in the West Indies, Haiti, Grenada, Barbados, Trinidad and Tobago, Antigua/Barbuda, Barbados, Belize, Jamaica, Montserrat, St. Kitts/Nevis, St. Vincent,
Rhipicephalus sanguineus	Widely distribution in the West Indies Anguilla, Antigua/Barbuda, Bahamas, Barbados, Bermuda, British Virgin Islands, Cayman Islands, Dominica, Grenada, Haiti, Jamaica, Montserrat, St.Kitts/Nevis, St. Lucia, St. Vincent, Turks and Caicos Islands, Trinidad and Tobago

transmitting diseases to ruminants in the Caribbean islands included *Amblyomma variegatum* (vector of cowdriosis and associated with acute dermatophilosis), *Amblyomma cajennense* (a potential vector of cowdriosis), and *R. (B). microplus* (a vector of babesiosis and anaplasmosis). The annual financial losses to the livestock industry due to tick infestation have been estimated to be in the millions of dollars per year in Commonwealth Caribbean countries alone.[88] It is believed that the most important tick in the Caribbean in terms of disease transmission, the tropical bont tick' *A. variegatum*, was first imported into the region by livestock coming from Senegal (West Africa) to the island of Guadeloupe (West Indies) around the 19th century (1828/30)[89] (http://entnemdept.ufl.edu/creatures/livestock/ticks/tropical_bont_tick.htm). It was speculated that these ticks later spread to other Caribbean islands.[90] A program with the primary objective to eradicate *A. variegatum* from nine islands (Anguilla, Antigua and Barbuda, Barbados, Dominica, Montserrat, Nevis, St. Kitts, St. Lucia, and St. Maarteen) was launched in 1994[91] and ended in 2008.[92] It has been estimated that the presence of *A. variegatum* in Guadeloupe means an annual economic loss for the 75,000 cattle on the island to be US$850,000.[93] Of this 50% of the losses was due to dermatophilosis, 17% to heartwater and 33% due to the physical damage caused by the ticks themselves. An additional US $700,000 was the estimated cost of acaricidal treatment.

CHAPTER 3

Tick Species Present in Trinidad and Tobago

Trinidad and Tobago is a water-locked country, located at Latitude 10.67 North and 61.52 West Longitude. This twin island state is bounded by the Caribbean Sea in the North, Gulf of Paria in the West, Atlantic Ocean in the East, and the Columbus Channel in the South. These islands lie approximately 11 km (6.8 miles) off the northeastern coast of Venezuela, from which they are separated by the Gulf of Paria. According to FAO (2013),[94] the estimated livestock population in Trinidad and Tobago was 34,500 cattle, 6100 buffaloes, 18,420 goats, 21,117 sheep, 35,000 pigs, 1400 horses, and 35,000,000 chickens.

A review of the literature on ticks of Trinidad and Tobago revealed 23 tick species, belonging to seven genera on the islands. The life cycle and control of most of these tick species, with the exception of *Amblyomma cajennense* and *Rhipicephalus (Boophilus) microplus*, have not been studied to date.[23,26,27,30,95] However, other available information including the distribution, associated pathogens, and effects on hosts of the 23 species distributed over seven genera have been recorded.[80] The problematic species will be emphasized in this book.

Of the seven genera recorded from Trinidad and Tobago, two belong to the family Argasidae with eight species. The remaining five genera belong to the family Ixodidae consisting of 15 species. These include nine species of the genus *Amblyomma*, six of *Ornithodoros*, two each of the genera *Rhipicephalus*, *Ixodes*, and *Argas*, and one each of the genera *Dermacentor* (*Anocentor*) and *Haemaphysalis*.[80]

The Argasid species recorded include:

- *Argas persicus*
- *Argas miniatus*
- *Ornithodoros azteci*
- *Ornithodoros capensis*
- *Ornithodoros denmarki*

Ticks of Trinidad and Tobago—An Overview. DOI: http://dx.doi.org/10.1016/B978-0-12-809744-1.00003-7

- *Ornithodoros hasei*
- *Ornithodoros puertoricensis*
- *Ornithodoros viguerasi*

The Ixodid species recorded include:

- *Amblyomma auricularium*
- *Amblyomma cajennense*
- *Amblyomma calcaratum*
- *Amblyomma dissimile*
- *Amblyomma humerale*
- *Amblyomma longirostre*
- *Amblyomma nodosum*
- *Amblyomma ovale*
- *Amblyomma rotundatum*
- *Dermacentor nitens*
- *Haemaphysalis juxtakochi*
- *Ixodes downsi*
- *Ixodes luciae*
- *Rhipicephalus (Boophilus) microplus*
- *Rhipicephalus sanguineus*

IDENTIFICATION CHARACTERS, DISTRIBUTION, HOSTS, AND DISEASE TRANSMISSION OF TICKS OF TRINIDAD AND TOBAGO

The Soft Ticks

1. *Argas (Persicargas) miniatus* Koch, 1844 (Fig. 3.1)

Identification characters[96]:

Larva of A. (P.) miniatus

- Oval idiosoma with striated integument
- Festoons absent
- No eyes
- Palps with four segments; medium-sized hypostome with heterodont teeth
- Dentition 3/3 in the anterior and 2/2 in the posterior portion

Distribution: Brazil, Guyana, Colombia, Panama, Venezuela, Puerto Rico, Cuba,[97] Mexico,[98] USA, Peru.[81]

Hosts: Fowls.

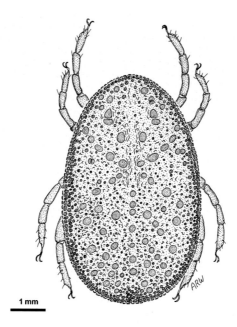

1 mm

Figure 3.1 Dorsal view of the female soft tick Argas (Persicargas) miniatus. [Figure from A.R. Walker, pers. comm.]

A. miniatus is a common parasite of chickens in Neotropical countries, which might have sometimes been misidentified as *A. persicus*. Many reports of *A. persicus*, a Palearctic species found on poultry in the Neotropics, may actually represent misidentifications of *A. miniatus* or another related species.[21]

Disease transmission: *A. miniatus* transmits *Borrelia anserina* (Borreliosis).[2]

Reports from Trinidad: *A. (P.) miniatus* has been detected in chickens from Gasparee Island, Trinidad.[21]

2. *Argas persicus* Oken, 1818 (Fig. 3.2)

Identification characters[36,99]:

- Oval body
- Males are slightly darker and smaller
- Lateral suture at the body margin marked by rectangular plates
- Leathery integument
- No eyes and dorsal shield
- Ventral capitulum
- Straight margin of basis capituli

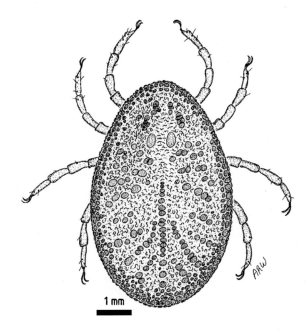

Figure 3.2 Dorsal view of the female soft tick Argas persicus. [Figure from Walker AR, Bouattour A, Camicas JJ, Estrada-pena A, Horak IG, Latif AA, et al. *Ticks of domestic animals in Africa: A guide to identification of species.* Bioscience Reports; 2003. p. 1–122].

- Articulation of palpi free
- No anal groove in both sexes
- No festoons and ventral plates in males
- Spiracular plate ovate and small

Distribution: Cosmopolitan,[81] especially in the tropics.

Hosts: Fowls, ducks, geese, turkeys, pigeons, canaries, and a variety of wild birds. This tick also bites humans.

As mentioned earlier, *A. persicus*, a Palearctic species, can be easily confounded with other *Argas* species during identification, and many reports of *A. persicus* on poultry in the Neotropics may actually represent misidentifications of *A. miniatus* or another related species.[21]

Disease transmission: *A. persicus* is known to transmit *B. anserina*, the cause of fowl spirochetosis (Borreliosis) and *Aegyptionella pullorum* (Aegyptianellosis) in domestic poultry.[2] The spirochetes may be transmitted from one generation of ticks to the next through the eggs and this transmission to the hosts occurs by biting or by fecal contamination.

The tick also transmits Slovakia virus.[100] Larval ticks produce a toxin, which causes paralysis in chickens[36] and in ducks. Reports of *A. persicus* parasitizing humans is rare.[101]

Reports from Trinidad: Hart[14] documented the presence of the fowl tick *A. persicus* and presumed that these ticks were imported from the United States to Trinidad and Tobago. He observed that the infested birds sat down, dropped their wings, and experienced fever. He also reported that dressing infested fowls with petroleum killed the ticks.

3. *Ornithodoros (Alectorobius) azteci* Matheson, 1935 (Fig. 3.3)

Identification characters[102,103]:

Larva:

- Possess 17–21 pairs of dorsal setae
- Hypostome bluntly pointed anteriorly
- Triangular to pyriform dorsal plate (length ~205 mm and width ~163 mm)
- Two pairs of basal setae on tarsus I

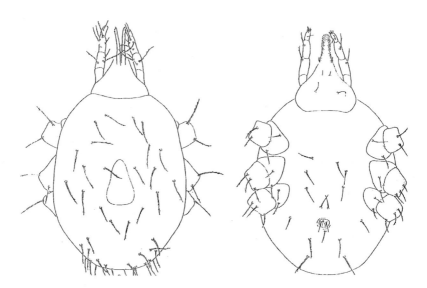

Figure 3.3 Dorsal and ventral views of the larva of soft tick Ornithodoros (Alectorobius) azteci. [Figure from Kohls GM, Sonenshine DE, Clifford CM. The systemics of the subfamily Ornithodorinae (Acarina: Argasidae). II. Identification of the larvae of the Western Hemisphere and descriptions of three new species. Ann Entomol Soc Am 1965;58(3):331–64].

Adult:

- Hump on tarsi I–IV
- Distinctly elevated mammillae in the preanal region
- Lack evident integument discs

Distribution: Colombia, Cuba, Jamaica, Islands of the Lesser Antilles, Mexico, Panama, Venezuela.[81,102]

Hosts: Bats (*Desmodus rotundus rotundus, Lonchorhina aurita, Peropteryx macrotis macrotis, Artibeus jamaicensis yucatanicus*).

Disease transmission: No recorded reports.

Reports from Trinidad: *O. (A.) azteci* has been collected from bats in Trinidad.[15,17,18] This species was also found on vampire bats (*D. rotundus rotundus*) from Croney Cave, Diego Martin in 1958 and from *L. aurita*, Paramin, Maraval, Trinidad, in 1960.[102]

4. *Ornithodoros (Alectorobius) capensis* Neumann, 1901 (Fig. 3.4)

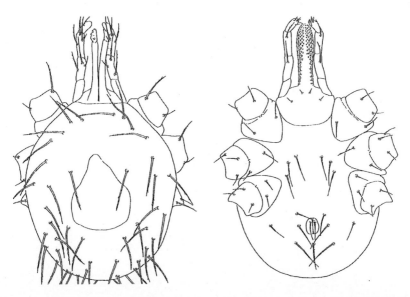

Figure 3.4 Dorsal and ventral views of the larva of soft tick Ornithodoros (Alectorobius) capensis. [Figure from Kohls GM, Sonenshine DE, Clifford CM. The systemics of the subfamily Ornithodorinae (Acarina: Argasidae). II. Identification of the larvae of the Western Hemisphere and descriptions of three new species. *Ann Entomol Soc Am* 1965;**58**(3):331–64].

Identification characters[102,104]:

Adult:

- Variable in size: females 1.70−3.5 mm wide and 3.0−6.5 mm long and males 1.4−3.0 mm wide and 2.5−5.0 mm long
- Sides of the tick are almost straight and posterior margin broadly rounded
- Discs are large and distinct on dorsum and arranged symmetrically
- On the venter, discs are smaller; mammillae are large, hemispherical, and glossy
- Legs: tarsus I has a mild subapical dorsal hump
- Capitulum sunken in a deep camerostome
- Basis capituli is approximately 1.4 times as wide as long
- Hypostome length is approximately 0.18 mm in male and 0.22 mm in female
- Teeth large and pointed. Arranged in one row of 3/3 and five rows of 2/2
- Eyes are absent
- Genital orifice situated at the posterior edge of the first pair of coxae

Larva:

- The length of unfed larvae is 0.635−0.748 mm and 0.377−0.440 mm wide including capitulum
- Dorsal plate is large, pyriform and widest posteriorly
- Dorsum with 22−25 pairs of setae (18−21 dorsolateral and 4 central)
- Capitulum is subtriangular dorsally and posterior margin weakly concave
- Basis capitulum is 0.138−0.166 mm and 0.177−0.210 mm wide
- Average length of palpal article 1 is 0.05 mm with no setae
- Article 2 is 0.068 mm with 4 setae
- Article 3 is 0.062 mm with 5 setae
- Article 4 is 0.047 mm with 8 or 9 setae
- Hypostome is elongated and arises from subtriangular median extension and tapers to a blunt apex
- Dentition is 5/5 in anterior third, 4/4 near midlength, and 2/2 posterior to the base

Nymph:

- Similar to adults except no aperture in the genital area of nymph

Hosts: Penguins, seabirds, shorebirds, coastal birds (cosmopolitan tick of seabirds[105]) and humans.[101]

Distribution: Dassen Island and Malagas Island (South Africa), Australia, Central Pacific, Ethiopia, Hawaii, Marshall Islands, New Zealand, Texas, Dominica, Jamaica, Cuba.[81,102]

Disease transmission: Between 1962 and 1965, 15 strains of Hughes virus were isolated from Soldado Rock, a small limestone island located about 9.6 km off the southwestern tip of Trinidad. Seven isolates came from ticks of the *O. capensis* complex and eight from nestling terns *Sterna fuscata*.[20] Jonkers et al.[106] reported Soldado virus (TRVL 52214), a new viral agent, isolated from a pool of nymphal *O. capensis* ticks (either *O. capensis sensu stricto* or *O. denmarki*) collected from a brown noddy (*Anous stolidus*) on Soldado Rock. *O. (A.) capensis* is also known to transmit other Soldado viruses in Hawaii, Seychelles, Trinidad, and Ethiopia,[100] Johnston Atoll virus, Abal virus in Central Pacific Islands, Australia, New Zealand, SW Africa,[107] Upolu virus in Australia, Baku virus in Azerbaijan, Aransas Bay virus in Texas, USA, Saumarez Reef virus in Tasmania, Hirota virus in Aomatsushima Island, Japan, and Midway virus in Hawaii.[100]

Chastel et al.[108] studied Soldado virus and indicated that *O. capensis* is a potential public health problem where sea gulls feed in urban habitats.

Reports from Trinidad: Kohls et al.[102] recorded *O. (A.) capensis* from spotted sandpipers (*Actitis macularia*) from Laventille swamp, Port of Spain, Trinidad. Aitken et al.[109] further recorded *O. (A.) capensis* during their survey of arthropods for natural virus infection.

5. *Ornithodoros* (*Alectorobius*) *denmarki* Kohls, Sonenshine, and Clifford, 1965 (Fig. 3.5)

Identification characters[102]:

Larva:

- The length of unfed larvae is 0.516−0.694 mm long and 0.342−0.469 mm wide including capitulum
- Dorsal plate is moderately large, pyriform, widest posteriorly
- Dorsum with 14−16 pairs of setae (11−13 dorsolateral and 3 central)
- Basis capitulum is 0.101−0.142 mm long and 0.152−0.196 mm wide
- Average length of palpal article 1 is 0.010 mm with no setae

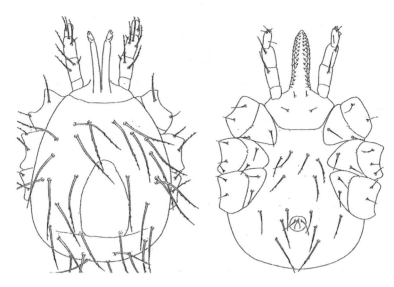

Figure 3.5 Dorsal and ventral views of the larva of soft tick Ornithodoros denmarki. [Figure from Kohls GM, Sonenshine DE, Clifford CM. The systemics of the subfamily Ornithodorinae (Acarina: Argasidae). II. Identification of the larvae of the Western Hemisphere and descriptions of three new species. *Ann Entomol Soc Am* 1965;**58**(3):331–64].

- Article 2 is 0.058 mm with 4 setae
- Article 3 is 0.060 mm with 5 setae
- Article 4 is 0.045 mm with 9 setae
- Hypostome arises from a small subtriangular median extension and tapers to a point (0.144–0.175 mm long and 0.049–0.060 mm wide)
- Dentition 3/3 in anterior half, 2/2 posteriorly base; file 1 with 16-18 denticles, file 2 with 14-17 and file 3 with 8-10
- Legs: Tarsus I is 0.189 – 0.223 mm long and 0.062 – 0.078 mm wide

Nymph:

- Oval shaped, pointed anteriorly
- Distinct, large dorsal disc & irregular in size and shape
- Distinct coxae I & II and other contiguous
- Spiracular plate situated between coxae III & IV
- Posteromedian seta absent

Adult:

- Variable in size: males 2.5 mm long and 1.2 mm wide and females 2.8 mm long and 1.5 mm wide

Hosts: Seabirds, sooty tern (*S. fuscata*) and noddy terns (*A. stolidus*).

Distribution: California, Hawaii, Florida, Ethiopia, Mexico, Jamaica, and Cuba.[97,102]

Disease transmission: *O. (A.) denmarki* transmits Hughes viruses in Florida, USA, NW Mexico, Venezuela, Cuba, and Trinidad; Farallon virus in California, Oregon, USA; Soldado in Hawaii, Seychelles, Ethiopia, and Trinidad; Raza virus in United States and Midway virus in Hawaii, Mexico.[100]

Reports from Trinidad: The larvae, nymphs, and adults of *O. (A.) denmarki*, a new species,[102] which was provisionally identified as *O. (A.) capensis* from Trinidad by Denmark and Clifford,[110] were found in nesting sites of sooty and noddy terns on Soldado Rock, Trinidad. Aitken et al.[109] recorded *O. (A.) denmarki*, during their survey of arthropods for natural virus infections.

6. *Ornithodoros (Alectorobius) hasei* Schulze, 1935 (Fig. 3.6)

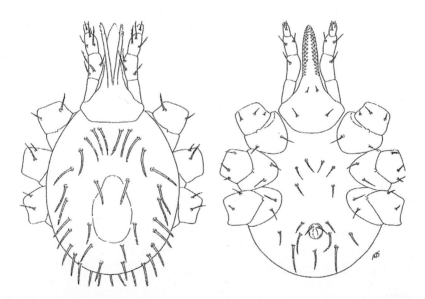

*Figure 3.6 Dorsal and ventral views of the larva of soft tick Ornithodoros (Alectorobius) hasei. [Figure from Kohls GM, Sonenshine DE, Clifford CM. The systemics of the subfamily Ornithodorinae (Acarina: Argasidae). II. Identification of the larvae of the Western Hemisphere and descriptions of three new species. Ann Entomol Soc Am 1965;**58**(3):331−64].*

Identification characters[102,111]:

Larva:

- Dorsal plate moderately large, pyriform, and narrow
- Hypostome arises from a small subtriangular median extension and taper to a point
- Apical dental formula: 3/3
- Median dental formula: 3/3
- Basal dental formula: 2/2
- Denticle row 1: 16−18 teeth
- Denticle row 2: 15−18 teeth
- Denticle row 3: 8−12 teeth

Distribution: Barbuda, Brazil, Guyana, Colombia, Costa Rica, Dominica, Guadeloupe, Guatemala, Martinique, Mexico, Nicaragua, Panama, Bolivia, Peru, Uruguay, St. Croix,[81] Venezuela, Bolivia,[102] Argentina,[111] Paraguay,[112] Jamaica.[97]

Hosts: Bats (*Noctilio leporinus leporinus, Molossops temminckii, Myotis albescens, Histiotus laephotis*).

Disease transmission: No recorded reports.

Reports from Trinidad: Kohls et al.[102] recorded *O. (A.) hasei* from bats (*N. leporinus leporinus*) in North Manzanilla, Trinidad.

7. *Ornithodoros (Alectorobius) puertoricensis* Fox, 1947 (Fig. 3.7)

Identification characters[102,113]:

Larva:

- Pyriform dorsal plate
- Long capitulum, length 0.120 mm from the posterior margin of the basis capituli to the posthypostomal setae
- Apically pointed hypostome
- Dental formula: 3/3 in the anterior half, 2/2 posteriorly almost to base
- Hypostomal length 0.232−0.266 mm, extremely long
- File 1 with 25−27 denticles, file 2 with 22−25, and file 3 with 14−16

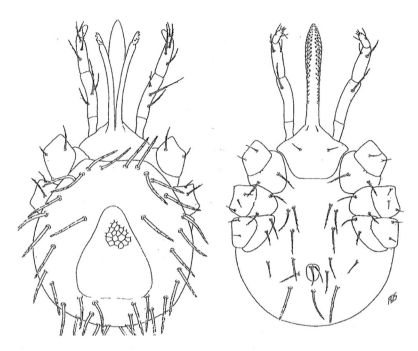

Figure 3.7 Dorsal and ventral views of the larva of soft tick Ornithodoros (Alectorobius) puertoricensis. [Figure from Kohls GM, Sonenshine DE, Clifford CM. The systemics of the subfamily Ornithodorinae (Acarina: Argasidae). II. Identification of the larvae of the Western Hemisphere and descriptions of three new species. *Ann Entomol Soc Am* 1965;**58**(3):331−64].

Distribution: Colombia, Jamaica, Panama, Virgin Islands, Puerto Rico, Guadeloupe, Nicaragua, Surinam, Uruguay, Argentina, Bolivia, Brazil, Paraguay, St. Croix, Venezuela.[102,81]

Hosts: This tick was first identified from rats in Puerto Rico.[114] The other hosts are humans,[102] domestic dogs,[115] mongoose, iguana, lizards, and other reptiles, (*Varanus dumerilii, Python regius,* and *Python bivittatus*).[113]

Disease transmission: O. *(A.) puertoricensis* is reported to transmit Asfavirus (African Swine Fever) to domestic pigs.[116]

Reports from Trinidad: Kohls et al.[102] recorded O. *(A.) puertoricensis* from the spiny rat (*Proechimys guyanensis trinitatus*) from Cumaca, Trinidad. Endris et al.[117] did a redescription of O. *(A.) puertoricensis* by scanning electron microscopy. They also reported host records for larvae of O. *(A.) puertoricensis* from rodents (*P. guyanensis trinitatus* and *Nectomys squamipes*) in Trinidad.

8. *Ornithodoros* (*Subparmatus*) *viguerasi* Cooley and Kohls, 1941 (Plate 3.1A−D)

Identification characters[102,118]:

Adult:

- A genital aperture covered by a semicircular flap
- A central sclerotized plate posterior to the genital aperture
- A transverse and thin plate anterior to the genital aperture located at the level of the anterior margin of coxae I

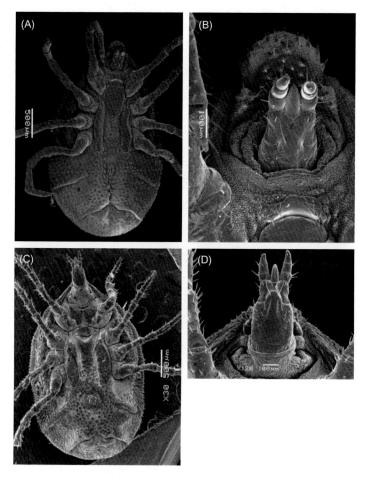

Plate 3.1 (A) Ventral view of Ornithodoros viguerasi, male.(B) Ventral view of the capitulum of Ornithodoros viguerasi, male. (C) Ventral view of Ornithodoros viguerasi female. (D) Ventral view of the capitulum of Ornithodoros viguerasi, female. [Photographs from Nava S, José MV, Enrique ARN, Atilio JM, Marcelo BL. Morphological study of Ornithodoros viguerasi Cooley and Kohls, 1941 (Acari: Ixodida: Argasidae), with sequence information from the mitochondrial 16S rDNA Gene. Acarologia 2012;52(1): 29−38].

- A pair of sclerotized plates bordering coxae II, III, and IV
- Basis capituli rectangular in shape and protrusible
- A hypostome roughly blunt at the apex with small denticles
- Article 1 of the palpi with a medial integumental extension, which has a long setae on the medial margin

Distribution: Cuba, Dominican Republic, Jamaica, Puerto Rico, Venezuela. Costa Rica, Haiti, Trinidad and Tobago, and Venezuela and Mexico.[81,97,118]

Hosts: Bats (*Pteronotus davyi, Pteronotus rosa rubiginosa, Mormoops megalophylla tumidiceps, Phyllonycteris poeyi, Pteronotus macleayii, Pteronotus quadridens, Pteronotus parnellii, Pteronotus gymnonotus, Brachyphylla nana, Mormoops megalophylla, Mormoops blainvillei, Erophylla sezekorni,* and *Erophylla bombifrons*).

Disease transmission: No recorded reports.

Reports from Trinidad: Kohls et al.[102] recorded *O. (S.) viguerasi* from *M. megalophylla tumidiceps* bats at Mount Tamana, and from *P. rosa rubiginosa* bats in Port of Spain, Tamana Hill cave, and Diego Martin. Aitken et al.[19] recorded *O. (S.) viguerasi* from the lesser naked-backed bat (*P. davyi*) from Bush Forest.

Hard Ticks

1. *Amblyomma auricularium* Conil, 1878 (Fig. 3.8A and B).

Identification characters[119,120]

- Lateral margin of basis capituli straight in male and convex in female
- Palpal segment I ventrally with a long triangular, rounded spur directed posteriorly and ventrally in male, whereas pyramidal spur directed ventrally in female
- Coxa with two rounded spurs, the external spur longer and more narrow
- Festoons narrow, large, and flat
- Dentition: 3/3

Hosts: Armadillos,[119,120] iguana, cattle, dogs, horses.[121]

Distribution: Ranges from Mexico to Argentina[119,120] and several ecoregions, Belize, Bolivia, Brazil, Colombia, Costa Rica, French Guiana, Honduras, Mexico, Nicaragua, Panama, Paraguay, Trinidad and Tobago, Uruguay, Venezuela[121] mainly in tropical and subtropical grasslands, savannas and shrublands, and broadleaf forests.

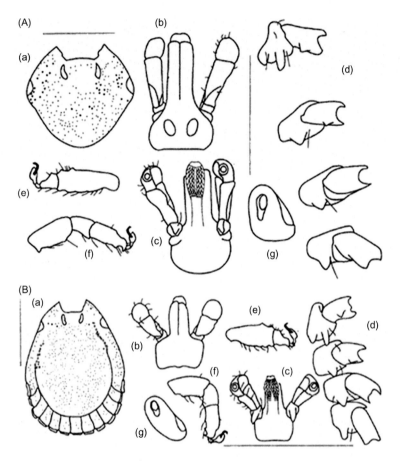

Figure 3.8 (A): Amblyomma auricularium, female: (a) scutum; (b) gnathosoma, dorsal view; (c) gnathosoma, ventral view; (d) coxae; (e) tarsus I; (f) tarsus and tibia; (g) spiracular plate. Scale bar 1 mm. (B): Amblyomma auricularium, male: (a) scutum; (b) gnathosoma, dorsal view; (c) gnathosoma, ventral view; (d) coxae; (e) tarsus I; (f) tarsus and tibia; (g) spiracular plate. Scale bar 1 mm. [Figure from Voltzit OV. A review of neotropical Amblyomma species (Acari: Ixodidae). Acarina 2007;15(1):3–134].

Disease transmission: No recorded reports.

Reports from Trinidad: Jones et al.[119] and Voltzit[120] mentioned the presence of *A. auricularium* in Trinidad.

2. *Amblyomma cajennense* Fabricius, 1787 (Plate 3.2A–D)

Identification characters[119,120]:

- Small to medium size tick (length 5 mm, width 3 mm)
- Scutum—ornamented, triangular, rounded anteriorly. Scutum with reddish brown spots in males and little broader than long in females

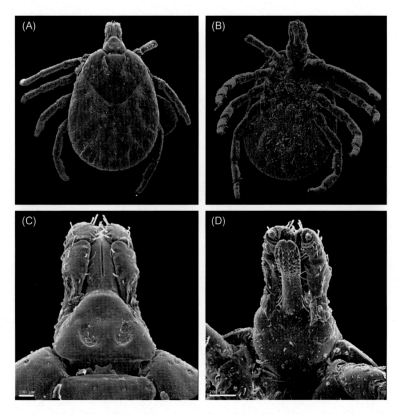

Plate 3.2 (A) Dorsal view of Amblyomma cajennense female. (B) Ventral view of Amblyomma cajennense female. (C) Dorsal view of the capitulum of Amblyomma cajennense female. (D) Ventral view of the capitulum of Amblyomma cajennense female.

- Coxa I with a pair of stout spurs—external is longer and pointed
- Coxa IV with a long pointed spur
- Festoons well defined
- Genital aperture at the level of the posterior margin of coxa II in males and opposite the interspace between coxae II and III in females
- Spiracles large and comma-shaped
- Basis capituli twice as broad as long
- Article 2 twice as long as article 3
- Porose areas small, circular, and well separated
- Hypostome dentition 3/3

Hosts: Cattle, sheep, goats, horses, donkeys, buffaloes, dogs, pigs, large rodent (*Hydrochoerus capybara*), giant toad (*Bufo marinus*), opossum, and humans.[13,23,24,26,27,122–124]

Distribution: Ranges from southern United States to northern Argentina and Caribbean Islands.[81,120,122]

Disease transmission: *A. cajennense* is known to transmit Wad Medani in Jamaica and *Rickettsia rickettsii* (Rocky Mountain spotted fever)[2] and *Leptospira pomona*. Newstead[125] recorded *A. cajennense* as the greatest pest in humans in Jamaica. Stoll[126] mentioned the annoyance caused to humans in Guatemala. All active stages of *A. cajennense* bite humans, leaving a painful lesion.[127] Parola et al.[128] reported the transmission of *Rickettsia honei* (Flinders island spotted fever) to humans by this tick. The tick may serve as a vector of Rocky Mountain spotted fever in western and central Mexico and South America.[129]

Reports from Trinidad: Neumann[13] is credited for the identification of *A. cajennense* from Trinidad. Aitken et al.[27] recorded "The 1958 Cayenne tick outbreak" at the Lagoon Doux Estate, south of Mayaro, where only the laborers showed serious effects of tick exposure, such as bouts of fever and skin irritations. These authors found *A. cajennense* on a variety of animals including humans. Spraying dichlorodiphenyltrichloroethane (DDT), Gamma Benzene Hexachloride (GBH), chlordane and dieldrin helped to control the tick population. Smith[23] found very large populations of *A. cajennense* ticks in infested areas with uncontrolled grass growth and suggested that a reduction in grass length and the removal of tree shade could help reduce the tick populations. He also studied the distribution of *A. cajennense* in Trinidad and Tobago; this tick was found only in Trinidad on the Cedros peninsula and on the east coast at Mayaro. Both places showed that annual rainfall was less than 175 cm with a marked dry season, well-drained sandy soils, and a permanent population of livestock. Smith[24] recorded *A. cajennense* in ruminants, equines, dogs and humans. The ecology and life cycle of the tick was investigated by Dindial[26] and Smith,[124] respectively. Dindial[26] found that *A. cajennense* was present in the coastal regions of Manzanilla, Mayaro, Guayaguauare, and the Cedros Peninsula area of Trinidad and was absent in Tobago. Clarkson[123] and Rawlins[130] in Trinidad also established the presence of *A. cajennense* on cattle. Lans[28] reported a second outbreak of this tick from 1994 to 1996 in Cedros and Mayaro linked to the presence of free-ranging cattle on the coconut estates.

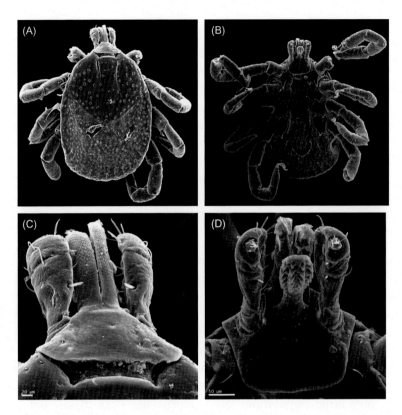

Plate 3.3 (A) Dorsal view of Amblyomma calcaratum female. (B) Ventral view of Amblyomma calcaratum female. (C) Dorsal view of the capitulum of Amblyomma calcaratum female. (D) Ventral view of the capitulum of Amblyomma calcaratum female.

3. *Amblyomma calcaratum* Neumann, 1899 (Plate 3.3A−D)

Identification characters[119,120]:

- Small tick
- Eyes large and flat
- Festoons longer than broad
- Genital aperture opposite coxa II
- Spiracles small, triangular with rounded angles in female and small, comma-shaped in male
- Scutum, cordiform in female and long oval in male
- Basis capituli—short, rectangular in male and triangular in female
- Article 2 longer than article 3
- Coxa I with a pair of long, pointed, subequal spurs
- Coxa II and III with a plate like spur
- Coxa IV with pointed spur
- Dentition: 3/3

Female Nymph:

- Medium size
- Scutum cordiform, ornate, irregular pale spots
- Article 2 of the palp very long
- Coxa I has two long spurs
- Coxae II—IV have one spur
- Eyes large
- Spiracle triangular, small with rounded angle
- Basis capituli triangular
- Hypostome dentition: 2/2

Hosts: Anteaters (*Tamandua longicaudata*), birds, small wild animals, dogs, humans.[19,24,119]

Distribution: Brazil, Bolivia, Colombia, Costa Rica, Belize, Guyana, Ecuador, Panama, Paraguay, Venezuela, Argentina, Peru.[119–121]

Disease transmission: No recorded reports.

Reports from Trinidad: *A. calcaratum* was recorded from Trinidad on anteaters, birds,[19] small wild animals and humans.[24]

4. *Amblyomma dissimile* Koch, 1844 (Plate 3.4A—D)

Identification characters[119,120]:

- Medium-sized tick; length 5 mm, width 4 mm
- Subtriangular scutum in females
- Ornamented scutum
- Eyes flat
- Hypostome dentition: 3/3
- Each coxa with two spurs, external spur longer than broad and internal spurs small
- Article 2 one and half times longer as article 3
- Genital aperture opposite coxa II
- Large spiracle and triangular
- Basis capituli rectangular and long in relation to breadth in males and broad wedge-shaped, rectangular in females
- Porose areas large, oval and distinct

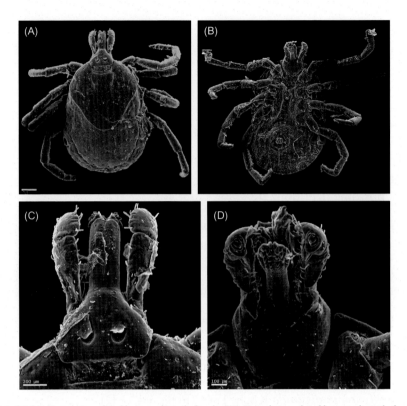

Plate 3.4 (A) Dorsal view of Amblyomma dissimile female. (B) Ventral view of Amblyomma dissimile female. (C) Dorsal view of the capitulum of Amblyomma dissimile female. (D) Ventral view of the capitulum of Amblyomma dissimile female.

Hosts: Toad (*B. marinus*), Snakes (*Boa constrictor, Lachesis mutus*), lizards (*Ameiva ameiva* and *Tupinambisnigro punctatus; iguana*), Caimans, small wild animals, cattle, humans.[1,19,24,109,120,122,131,132]

Distribution: Florida, Mexico, Georgia, West Indies to Argentina.[81,120]

Disease transmission: No recorded reports.

Reports from Trinidad: Neumann[13] is credited for the identification of *A. dissimile* in Trinidad. Nuttall et al.[122] recorded this tick in Trinidad from snakes (*Lachesis mutans*) and iguanas. Turk[131] found *Amblyomma trinitatis*, a synonym of *A. dissimile*, on a ground lizard at St. Augustine. Aitken et al.[19,109] recorded *A. dissimile*, during their survey of arthropods for natural virus infection and found it on various species of snake, large lizards, caimans, large toads and tortoises. Dindial[26] reported the presence of *A. dissimile* from iguanas, frogs and snakes both in Trinidad and Tobago. Voltzit[120] also mentioned the presence of *A. dissimile*, which feeds on reptiles and toads.

5. *Amblyomma humerale* Koch, 1844 (Plate 3.5A−C)

Identification characters[122]:

- Medium-sized tick
- In male, scutum smooth, ornate with small irregular spots and in female, scutum triangular, cordiform, and ornate
- Marginal groove absent
- Legs are long, coxae I−IV each with two short, blunt spurs, tarsi long
- Eyes present
- Punctuations numerous and irregular
- Large festoons

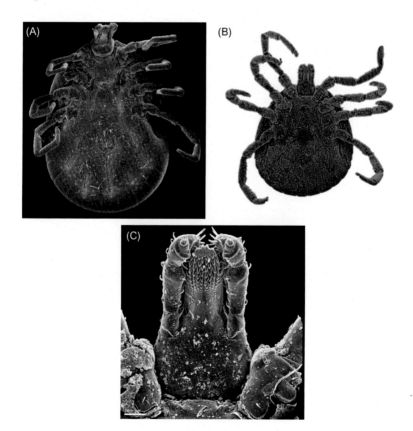

Plate 3.5 (A) Ventral view of Amblyomma humerale female. (B) Ventral view of Amblyomma humerale male. (C) Ventral view of the capitulum of Amblyomma humerale male.

- Genital aperture opposite to coxa II in male and between coxa II and III in female
- Article 2 more than double as long as article 3
- Spiracle large, comma-shaped in male and triangular with rounded angle in female
- Hypostome long, spatulate
- Dentition: 4/4

Hosts: Tortoise (*Geochelone denticulata*), reptiles, rodents (*Oryzomys capito* and *Proechimys guyannensis*).[19,22,120,133] Other observed hosts include birds,[134] lizards, opossums, anteaters,[135] and dogs.[136]

Distribution: Brazil, Guyana, Venezuela, Colombia, Ecuador Suriname, Peru, Trinidad and Tobago, Venezuela, Bolivia, and French Guiana.[97,120]

Disease transmission: No recorded reports.

Reports from Trinidad: *A. humerale* has been collected from the yellow-footed tortoise (*G. denticulata*) from Mayaro.[19,133] Everard and Tikasingh[22] found *A. humerale* on rodents (*O. capito* and *P. guyannensis*) from Turure Forest, Trinidad. Nava et al.[112] and Voltzit[120] also mentioned the presence of *A. humerale* in Trinidad.

6. *Amblyomma longirostre* Koch, 1844 (Plate 3.6A−C)

Identification characters[122]:

- Large tick—7 mm long, 4 mm wide
- Body elongate, oval and narrow anteriorly
- Ornate
- Scutum elongated in male and elongated oval in female
- Festoons well defined
- Capitulum: Hexagonal
- Palp long, club-shaped
- Eyes pale and flat
- In males ventral plaques large and median, lateral plaques elongate
- Genital aperture opposite coxa II
- Spiracles large, comma-shaped in male and triangular with rounded angle in female
- Article 2 twice as long as article 3
- Hypostome narrow and spatulate in male and very long, lanceolate in female

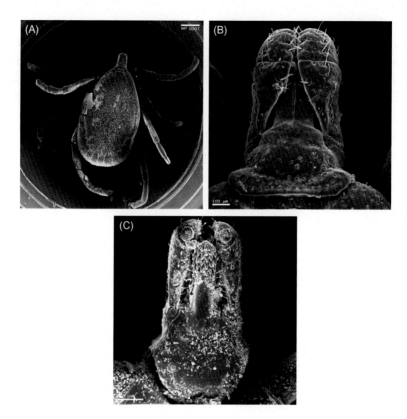

Plate 3.6 (A) Dorsal view of Amblyomma longirostre male nymph. (B) Dorsal view of the capitulum of Amblyomma longirostre male nymph. (C) Ventral view of the capitulum of Amblyomma longirostre male nymph.

- Dentition: 3/3
- Leg: very long; coxa I with two unequal, short spurs; a single short, pointed spur on each of coxae II–IV, internal spur small
- Porose areas small and widely separated in female

Hosts: Birds, porcupines, small wild animals,[19,24,109] and humans.[137]

Distribution: Bolivia, Brazil, French Guiana, Paraguay, Trinidad and Tobago, Venezuela, Colombia, Panama, southern Mexico, Uruguay, and Argentina.[97,119,138]

Disease transmission: *A. longirostre* collected from a Brazilian porcupine were found to be infected with a rickettsial strain of unknown pathogenicity for humans.[139]

Reports from Trinidad: Nuttall et al.[122] recorded this tick from Trinidad. Aitken et al.[19,109] recorded *A. longirostre* during their survey of arthropods for natural virus infection and its immature stages frequently parasitized birds. Everard and Tikasingh[22] found this tick on rodents (*O. capito* and *P. guyannensis*) from Turure Forest. Smith[24] recorded *A. longirostre* in Trinidad from small wild animals. Nava et al.[112] also indicated the presence of this tick in Trinidad. Voltzit[120] mentioned that in 1955 T.H.G. Aitken collected two males and a female respectively from porcupines (*Coendou prehensilis*) and birds (*Turdus nudigensis*) at Sangre Grande, Trinidad.

7. *Amblyomma nodosum* Neumann, 1899 (Fig. 3.9A and B)

Identification characters[120,122]:

- Small tick; length ∼5 mm, width ∼3.5 mm
- Capitulum broad and short in males and medium sizes in females
- Scutum: slightly convex with pale ornamentation in males and oval cordiform in females
- Eyes pale and flat
- Cervical grooves: short, deep, oval pits and very divergent
- Festoons very distinct/clearly defined and longer than broad
- Genital aperture opposite to coxa II
- Spiracle small short, pear/comma-shaped, broader in female
- Medium size of porose areas
- Hypostome long in females and short, broad and spatulate in males
- Palps short and conical
- Dentition: 3/3
- Leg: In males, coxa I with two long spurs, coxa II and III each with a short spur, coxa IV a single long spur. In females coxa I with two subequal spurs, coxae II, III, and IV with a single short, broad/flattened spur
- Tarsi less abruptly attenuated than with male and gorged in female

Hosts: Birds and anteaters (*Myrmecophaga jubata* and *Tamandua tetradactyla*).[112,119]

Distribution: Brazil, Bolivia, Costa Rica, Colombia, Guatemala, Nicaragua, Panama, Trinidad and Tobago, Venezuela, Argentina, Mexico, and Paraguay.[97,112,119,120,140]

Disease transmission: This tick transmits rickettsial infections.[140]

Reports from Trinidad: The presence of *A. nodosum* was recorded in Trinidad by several authors.[112,119,120]

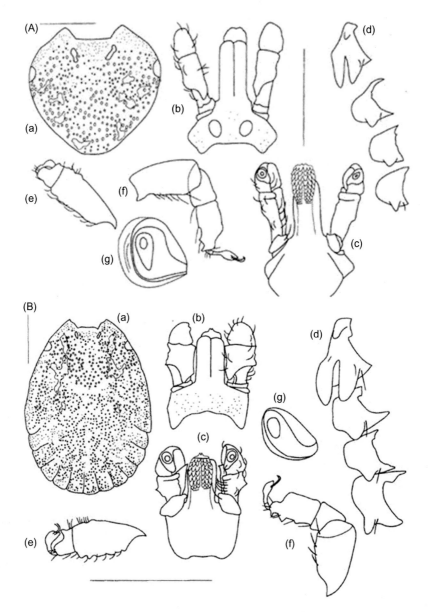

Figure 3.9 (A): Amblyomma nodosum, female: (a) scutum; (b) gnathosoma, dorsal view; (c) gnathosoma, ventral view; (d) coxae; (e) tarsus I; (f) tarsus and tibia; (g) spiracular plate. Scale bar 1 mm. (B): Amblyomma nodosum, male: (a) scutum; (b) gnathosoma, dorsal view; (c) gnathosoma, ventral view; (d) coxae; (e) tarsus I; (f) tarsus and tibia; (g) spiracular plate. Scale bar 1 mm. [Figure from Voltzit OV. A review of neotropical *Amblyomma* species (Acari: Ixodidae). *Acarina* 2007;**15**(1):3−134].

8. *Amblyomma ovale* Koch, 1844 (Fig. 3.10A and B)

Identification characters[122]:

- Small ornate tick, ~4 mm long, ~3 mm wide
- Sutum elongate and oval in males; triangular and cordiform in females
- Basis capituli subtriangular in female
- Hypostome dentition: 3/3
- Coxa I with two long, pointed contiguous spurs. Broad and salient ridge on coxa II and coxa III. Long spur on coxa IV in male and short spur on female
- Eyes large and flat
- Festoons well defined
- Genital aperture opposite coxa II in male and opposite the interspace between coxae II and III in female
- Spiracle large and comma-shaped in males and large, triangular in females

Hosts: Dogs, wolves, the aguarachay (*Canis azarae*), deer (*Cervus* sp.), the greater grison (*Galictis vittata*) and humans.[24,97,101,122,141]

Distribution: Mexico to Argentina[119,120] including Caribbean islands.[22,24,26]

Disease transmission: It carries *Hepatozoon* spp. in Brazil.[142]

Reports from Trinidad: Everard and Tikasingh[22] and Smith[24] recorded *A. ovale* from dogs. Dindial[26] found *A. ovale* associated with hunting dogs at Sans Souci and Charlotteville, Trinidad.

9. *Amblyomma rotundatum* Koch, 1844 (Fig. 3.11A and B)

Identification characters[122]:

- Small tick ~3.5 mm long and ~3 mm wide, oval
- Ornate
- Scutum oval in males and cordiform in females
- Marginal groove distinct
- Capitulum long
- Eyes situated at the anterior third of the scutum
- Coxae I–IV each with two short, rounded spurs

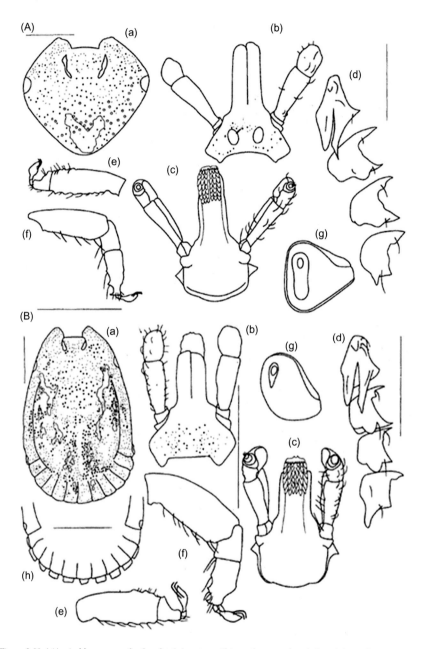

Figure 3.10 (A): Amblyomma ovale, female: (a) scutum; (b) gnathosoma, dorsal view; (c) gnathosoma, ventral view; (d) coxae; (e) tarsus I; (f) tarsus and tibia; (g) spiracular plate, Scale bar 1 mm. (B) Amblyomma ovale male: (a) scutum; (b) gnathosoma, dorsal view; (c) gnathosoma, ventral view; (d) coxae; (e) tarsus I; (f) tarsus and tibia; (g) spiracular plate; (h) posterior part of idiosoma, ventral view. Scale bar 1 mm. [Figure from Voltzit OV. A review of neotropical *Amblyomma* species (Acari: Ixodidae). *Acarina* 2007;**15**(1):3–134].

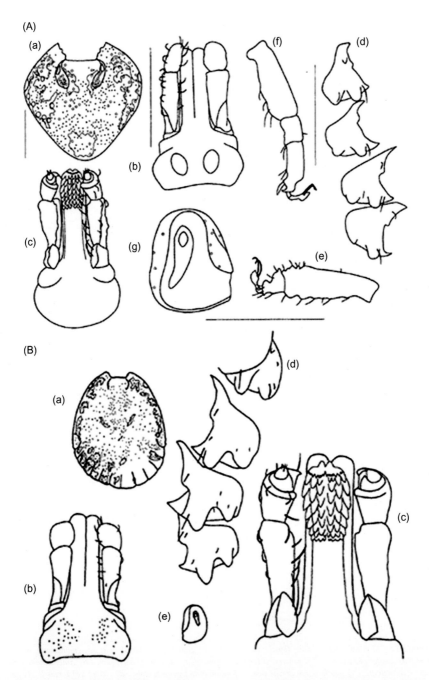

*Figure 3.11 (A):Amblyomma rotundatum, female: (a) scutum; (b) gnathosoma, dorsal view; (c) gnathosoma, ventral view; (d) coxae; (e) tarsus I; (f) tarsus and tibia; (g) spiracular plate. Scale bar 1 mm. (B): Amblyomma rotundatum, male: (a) scutum; (b) gnathosoma, dorsal view; (c) gnathosoma, ventral view; (d) coxae; (e) spiracular plate. [Figure from Voltzit OV. A review of neotropical Amblyomma species (Acari: Ixodidae). Acarina 2007;**15**(1):3−134].*

- Tarsi attenuated in talus
- Genital aperture situated opposite coxa II
- Spiracles ovoid-triangular
- Festoons present
- Porose areas medium in size, ovoid and widely separated
- Article 2 twice as long as article 3
- Hypostome dentition: 3/3

Hosts: Toads,[119] iguana, snakes, and humans.[143]

Distribution: Ranges from southern United States to northern Argentina and Caribbean Islands.[119,120,143]

Disease transmission: No recorded reports.

Reports from Trinidad: Jones et al.[119] stated that *A. rotundatum*, exclusively a parasite of cold-blooded animals,[144] was found to exploit toads in Trinidad.

10. *Dermacentor nitens* Neumann, 1897 (Plate 3.7A−E)

Identification characters[35]:

- Spiracle goblets form a ring
- Coxae I have larger and equal paired spurs
- Well-developed spurs of coxae II−IV
- Anal groove is indistinct
- Palpal segment 4 terminal
- Hypostomal dentition: 4/4

Hosts: Horses, donkeys[24,26], humans.[145] Guglielmone et al.[146] reported some exceptional hosts like toads and snakes. Smith[24] and Rawlins[130] recorded *D. nitens* from equines in Trinidad.

Distribution: Throughout the Caribbean and Galapagos Islands, Central America, Bolivia, Brazil.[35,81]

Disease transmission: *D. nitens* is known to transmit *Babesia caballi* (Equine babesiosis).[2] Asgarali et al.[31] investigated 93 horses in Trinidad for serum antibodies to *Theileria equi* and *B. caballi* using an immunofluorescent antibody test and found 77 to be seropositive. Very recently, Georges[32] diagnosed some tick-transmitted hemopathogens

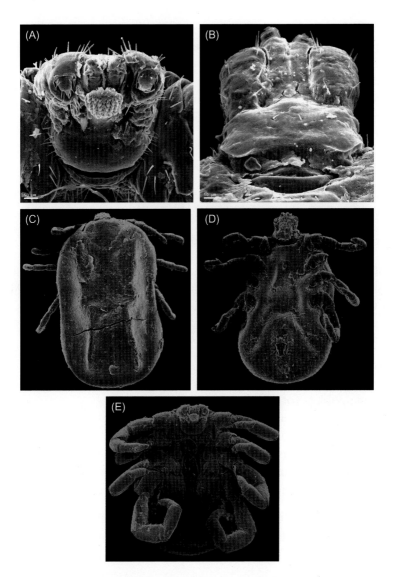

Plate 3.7 (A) Ventral view of the capitulum of Dermacentor nitens, male. (B) Dorsal view of the capitulum of Dermacentor nitens, male. (C) Dorsal view of Dermacentor nitens, female. (D) Ventral view of Dermacentor nitens, female. (E) Ventral view of Dermacentor nitens, male.

(*Anaplasma platys*, *Babesia canis vogeli*, *B. caballi*, *T. equi*) in companion animals using molecular tools.

Reports from Trinidad: Floch and Fauman[144] confirmed the presence of the tick, *D. nitens*, in Trinidad. The authors have also observed *D. nitens* on horses in Trinidad.

11. *Haemaphysalis juxtakochi* Cooley, 1946 (Fig. 3.12)

Identification characters[147]:

- Small, ~3 mm long, and ~2 mm wide; oval
- Marginal grooves distinct in females
- Scutum occupies half the length of the body in females
- Cornua as long as wide and pointed
- Faint porose areas
- Dentition: 4/4
- Trochanter I with distinct dorsal trochantal spur and absent on II–IV and mild ventral trochantal spurs present on I, II, and III

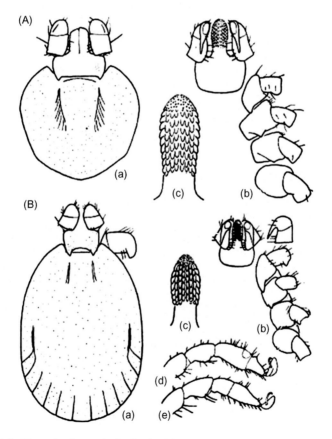

Figure 3.12 (A): Haemaphysalis juxtakochi, female: (a) capitulum and scutum, dorsum. (b) capitulum and coxae, ventrum. (c) hypostome. (B) Haemaphysalis juxtakochi, male: (a) capitulum and scutum, dorsum. (b) capitulum and coxae, ventrum, and side view of palpus. (c) hypostome. (d) metatarsus and tarsus, leg I. (e) metatarsus and tarsus. [Figure from Cooley, R.A. (1946) The genera Boophilus, Rhipicephalus, and Haemaphysalis (Ixodoidea) of the New World. Bulletin of the National Institute of Health 1946;(187): 1–54].

- Absent ventral spur on tarsi
- Genital aperture situated posterior between coxae II and III in females and between coxa II in males.

Hosts: Deer (*Mazama rufa, Mazama gouazoubira, Axis axis*); Paca (*Coelogenys paca*); wild boar (*Sus scrofa*).[26,148,149]

Distribution: Ranges from United States to northern Argentina.[97]

Disease transmission: *Rickettsia rhipicephali* and *Rickettsia bellii* were isolated from *H. juxtakochi* ticks in the State of São Paulo, Brazil.[150]

Reports from Trinidad: Kohls[148] first recorded the identity of *H. juxtakochi*, collected by T.H.G. Aitken on June 24, 1954 from Cumaca, Trinidad. This tick parasitizes deer (*M. rufa*) and paca (*C. paca*). Keirans[132] also recorded this tick from Caparo, Trinidad.

12. *Ixodes downsi* Kohls, 1957 (Plate 3.8A−D)

Identification characters[17,151]:

Adult:

- Length: 2.57−3.07 mm (unfed), tips of scapulae to posterior margin of body
- Width: 1.87−2.15 mm.
- Body: suboval, wider anteriorly in male and wider posteriorly in female
- Color: Males yellow brown with paler legs; females pale yellow with capitulum, scutum and legs yellow brown.
- Cornua absent in males but short and blunt in females
- Porose areas large, broader than long and shallow
- Auriculae absent in males but short and blunt in females
- Hypostome shorter than the palpi, broad, notched apically
- Denticles arranged 3/3 apically, then 2/2
- Scapulae short, blunt
- Coxae: in males, all coxae with a short external spur; in females, a broad external spur on coxa I, a similar but shorter spur on coxae II, III, and IV; a minute internal spur on coxa I in males but internal spurs absent in females
- Trochanter spurs absent

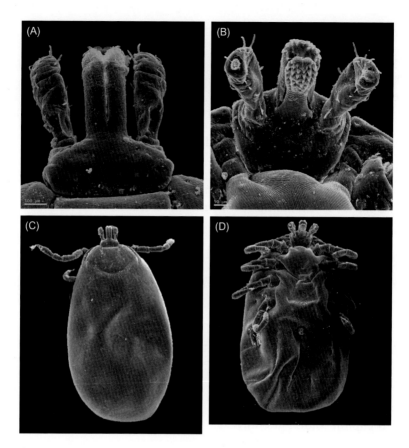

Plate 3.8 (A) Dorsal view of the capitulum of Ixodes downsi, female. (B) Ventral view of the capitulum of Ixodes downsi, female. (C) Dorsal view of Ixodes downsi, female. (D) Ventral view of Ixodes downsi, female.

- Spiracular plate: subcircular/suboval, goblets numerous and small
- Anal groove rounded in front of the anus, a little convergent posteriorly
- Genital aperture situated between coxae III

Nymph:

- Capitulum length 0.35 mm
- Hypostome as long as the palpi
- Dentition: 3/3 apically, then 2/2
- Scapulae are very short and blunt
- Scutum 0.57−0.59 mm long and 0.52−0.57 mm wide
- Spiracular plate-subelliptical, goblets fewer and larger than in the females

Larva:

- Hypostome as long as the palpi
- Principal dentition 2/2
- Cervical grooves distinct, shallow and divergent
- Coxae have short triangular external spurs and no internal spurs.

Distribution: Venezuela[151]; Trinidad[17]; Peru.[152]

Hosts: Bat (*Anoura geoffroyi*).[17] Gonzalez-acuna et al.[151] collected the tick from Guacharos cave, Venezuela, inhabited by bats and oilbird (Guacharo—*Steatornis caripensis*). No human infestations have been recorded.[146]

Disease transmission: No recoded reports.

Reports from Trinidad: Kohls[16] first recorded a new species of tick, *I. downsi* on bats (*A. geoffroyi*) in el Cerro del Aripo (10.72N 61.25W), Trinidad.

13. *Ixodes luciae* Sénevet, 1940 (Plate 3.9A–D)

Identification characters[153]:

- Cornua and auriculae absent
- Basis capituli flared laterally
- Hypostome smaller in males
- Dental formula 2/2
- Coxa I robust and bifid, posterior margin between internal and external spurs curved
- External spur on coxa I much longer than internal spur
- Coxae II–IV with large external spurs, rounded apically
- Trochanters I–III without spurs

Hosts: Rodents (*Oryzomys capito* (*syn. Hylaeamys megacephalus*), *O. laticeps*, *O. concolor*, *P. guyannensis*, *Zygodontomys*, *Heteromys*), Marsupials (*Didelphis marsupialis* and *Marmosa murina*, *Philander opossum*, *Caluromys lanatus*, *Marmosops* sp., *Metachirus nudicaudatus*, *Micoureus* sp., *P. andersoni*), small wild animals.[18,19,24,26,109,154]

Distribution: Argentina, Brazil, Bolivia, Colombia, Guyana, Guatemala, Peru, Mexico, Nicaragua, Surinam, Venezuela, Belize, Costa Rica, Ecuador, French Guiana, Panama, and Brazil.[155,156]

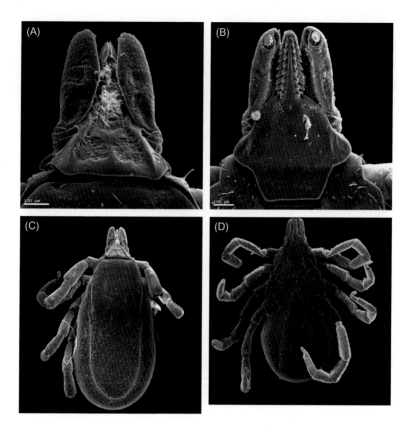

Plate 3.9 (A) Dorsal view of the capitulum of Ixodes luciae, female. (B) Ventral view of the capitulum of Ixodes luciae, female. (C) Dorsal view of Ixodes luciae, female. (D) Ventral view of Ixodes luciae, female.

Disease transmission: No recorded reports.

Reports from Trinidad: Aitken et al.[109] recorded *I. luciae* during their survey of arthropods for natural viral infections. Everard and Tikasingh[22] found this tick on rodents (*O. capito* and *P. guyannensis*) from Turure Forest. In 1974, Smith recorded *I. luciae* from small wild animals from Trinidad.

14. *Rhipicephalus (Boophilus) microplus* Canestrini, 1888 (Plate 3.10A−C)

Identification characters[36]:

- Dentition: 4/4
- Porose areas broad oval
- Cornua distinct in males

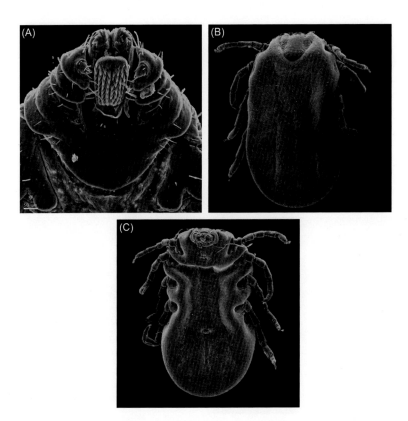

Plate 3.10 (A) Ventral view of the capitulum of Rhipicephalus (Boophilus) microplus, female. (B) Dorsal view of Rhipicephalus (Boophilus) microplus, female. (C) Ventral view of Rhipicephalus (Boophilus) microplus, female.

- In females palp articles 1 short, concave and internal margin without any protuberance
- Coxae I with distinct spurs in females and long in males
- U-shaped posterior lip of genital aperture
- Presence of caudal appendage in males
- Four indistinct ventral plate spurs (adanal plates), not visible dorsally

Distribution: Cosmopolitan; chiefly in tropical and subtropical areas worldwide.

Hosts: Cattle,   other   livestock   and   wildlife animals.[19,23,24,26,95,109,123,145] This tick has been reported on humans from Cuba,[157] Argentina,[158] and China.[159]

Disease transmission: *R. (B.) microplus* is known to transmit numerous parasites: *Babesia bovis* and *Babesia bigemina* (Bovine babesiosis),

Anaplasma marginale (bovine anaplasmosis) and *T. equi* (equine piroplasmosis).[2] A virus, Wad Medani has also been reported to be transmitted by *R. (B.) microplus* in Singapore and Malaysia.[100] There are no diseases known to be transmitted by *Boophilus* ticks to man.[101]

Reports from Trinidad: Aitken et al.[27] recorded a few specimens of the southern cattle tick *R. (B.) microplus* from cattle and sheep in Trinidad. Williams and Gonzalez[160] observed that Holstein heifers imported into Trinidad from Canada with the intent to develop a dairy industry were exposed to tick-infested pastures and that these exotic cattle suffered from a febrile disease associated with marked anemia, occasional hemoglobinuria and sometimes death within 2–6 weeks after importation and introduction to tick-infested pastures. They also confirmed the detection of *Babesia* spp. and *A. marginale* from the blood of the infected cattle. This established the presence of *R. (B.) microplus* ticks in Trinidad, as these ticks are vectors of both the pathogens. The presence of *R. (B.) microplus* on cattle in Trinidad was confirmed by Aitken et al.,[109] Clarkson,[123] and Rawlins.[130] Smith[23] studied the distribution of this tick in Trinidad and Tobago. *R. (B.) microplus* can be found throughout both islands, except on land newly cleared from forest. Furthermore, Smith[24] recorded this tick exploiting ruminants and equines. The prevalence and biology of *R. (B.) microplus* was investigated by Dindial[26] and he also reported the presence of this tick all over the country. Spraying DDT, GBH, chlordane and dieldrin was reported to have helped to control this tick population in Trinidad. Polar and coinvestigators[29] reported the use of entomopathogenic fungi *Metarhizium anisopliae* to control *R. (B.) microplus* at Aripo livestock station in Trinidad. Polar[30] found *M. anisopliae* to be effective against all developmental stages of *R. (B.) microplus* and *R. sanguineus*, except larvae of the latter.

15. *Rhipicephalus sanguineus* Latreille, 1806 (Plate 3.11A–D)

Identification characters[161]:

- Small, inornate with slight sexual dimorphism
- Elongated body
- Short palps
- Eyes and festoons are present
- Coxa I deeply clefted

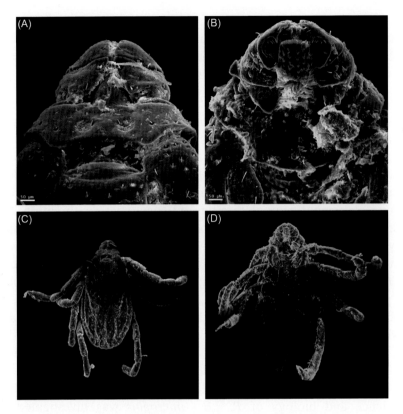

Plate 3.11 (A) Dorsal view of the capitulum of Rhipicephalus sanguineus, female. (B) Ventral view of the capitulum of Rhipicephalus sanguineus, female. (C) Dorsal view of Rhipicephalus sanguineus, female. (D) ventral view of Rhipicephalus sanguineus, female.

- Spiracular plates are comma-shaped in males
- Hexagonal basis capituli

Distribution: Cosmopolitan.

Host: Dogs,[19,24,26,109] wild and domestic animals, including humans.[101,162]

Disease transmission: *R. sanguineus* transmits a number of tick borne diseases and pathogens to their mammalian hosts. They include Wad Medani virus, *A. platys, Ehrlichia canis* (canine ehrlichiosis), *Babesia vogeli, Babesia canis, Babesia gibsoni* (canine babesiosis), *Hepatozoon canis* (canine hepatozoonosis), *R. rickettsii* (Rocky Mountain spotted fever), *Rickettsia conorii conorii* (tick bite fever/ Mediterranean spotted fever), and *Acanthocheilonema dracunculoides* (nematode) in dogs. *R. sanguineus* transmit *B. bigemina* (cattle babesiosis) in cattle and a number of diseases/pathogens in humans. They include Wad Medani virus, *R. rickettsii* (Rocky Mountain spotted fever), *R. conorii conorii* (tick bite fever/Mediterranean spotted fever), *Rickettsia conorii israelensis* (Israeli spotted fever), *R. conorii caspia* (Astrakhan fever), *Rickettsia conorii indica* (Indian tick fever), *Rickettsia massiliae* (unnamed), and *Coxiella burnetii* (Acute Q fever).[2,4,100,128,161]

Reports from Trinidad: Aitken et al.[109] recorded *R. sanguineus* during their survey of arthropods for natural virus infection. Smith[24] documented *R. sanguineus* from dogs.

CONCLUSION

Trinidad and Tobago harbor a high diversity of tick species, some of which play a significant role in animal and human health. The reported detection of 23 species of ticks is very high considering the relatively small size of the twin island state (Trinidad—4828 km^2; Tobago—300 km^2). Tobago, being the smaller of the two islands, has fewer animal populations and recorded lower tick species richness. *Amblyomma cajennense*, the main species of concern to humans and animals, is found only in some pockets of Trinidad and has not been reported from Tobago. The main factor limiting the distribution of this tick appears to be its intolerance of wet soil and high rainfall. The relatively small areas infested with this species have common features such as deep sandy, well-drained soil with a well-marked dry season and low rainfall.[23]

All reported surveys of ticks in the Caribbean suggest the absence of *Antricola mexicanus* and *Amblyomma variegatum* in Trinidad and Tobago, except an unpublished report[163] and an unconfirmed record by Rawlins et al.,[84] which was published in a newsletter. Alderink and McCauley[164] mentioned Trinidad and Tobago as one of the 12 islands non-infested by *A. variegatum* in the Lesser Antilles. The absence of *A. variegatum* in Trinidad and Tobago may be due to strict vigilance during animal importation. In addition, according to the annual report on Animal Health 2010, cowdriosis is absent in Trinidad and Tobago.[12] Cowdriosis is caused by the pathogen *Ehrlichia ruminantium*, which is transmitted by *Amblyomma* species. Among the *Amblyomma* species transmitting this pathogen, *A. cajennense* and *A. dissimile* are present in Trinidad and Tobago. Both these species are experimental and potential vectors of cowdriosis, and are thought to be poor vectors of this disease, as successful transmission by these ticks seem to be low.[165,166] Moreover, *A. cajennense* has not been incriminated thus far in the natural transmission of this disease (http://www.oie.int). *A. variegatum* is a natural vector of cowdriosis. So the absence of cowdriosis in Trinidad and Tobago leads us to conclude that *A. variegatum* may be absent in this country.

Hunter[167] reported the presence of *Rhipicephalus (Boophilus) annulatus* in Trinidad. However, Smith[24] concluded that the presence of this species was probably an incorrect identification of *Rhipicephalus (Boophilus) microplus*, a species found in abundance throughout Trinidad and Tobago. Subsequently, there has been no record of the presence of *R. (B.) annulatus* from Trinidad. It has been observed that the distribution of *R. (B.) microplus* on ruminants and *R. sanguineus* in dogs was very common in Trinidad and Tobago. Additionally, the temperature and humidity of Trinidad are very conducive for the multiplication of ticks year round. In controlling tick populations especially *R. (B.) microplus*, treatment programmes with acaricides should be undertaken twice a year- at the beginning and at the end of the wet season. A project can also be undertaken, by using a vaccine to eliminate *R. (B.) microplus* since encouraging results have been found in Cuba, Australia and other countries using this method.

Since 1977, there has been no comprehensive study on ticks in Trinidad and Tobago, and only sporadic information is available on disease transmission. Future studies on the ecology of ticks and epidemiology of tick-borne diseases in Trinidad and Tobago will therefore be rewarding, and veterinarians as well as researchers should be encouraged to update the list of prevalent ticks as this will lead to a better understanding of the epidemiology of tick-borne human and animal diseases in the country.

REFERENCES

1. Guglielmone AA, Robbins RG, Apanaskevich DA, Petney TN, Estrada-Pena A, Horak IG, et al. The Argasidae, Ixodidae and Nuttalliellidae (Acari: Ixodida) of the world: a list of valid species names. *Zootaxa* 2010;**2528**:1−28.

2. Jongejan F, Uilenberg G. The global importance of ticks. *Parasitology* 2004;**129**:S3−14.

3. Kaufman WR. Ticks: physiology aspects with implications for pathogen transmission. *Ticks Tick Borne Dis* 2010;**1**:11−22.

4. Parola P, Raoult D. Ticks and tickborne bacterial diseases in humans: an emerging infectious threat. *Clin Infect Dis* 2001;**32**:897−928.

5. Sonenshine DE. *Biology of ticks*, vol. 2. Oxford: Oxford University Press; 1993.

6. Nijhof A, Guglielmone A, Horak I, Latif A, Bouattour A, Ahmed J, et al. A global tick-host-pathogen database (THPbase). *Newsletter on Ticks and Tick-Borne Diseases of Livestock in the Tropics (ICTTD-3)* 2005;**28**:3.

7. FAO. *Ticks and tick borne disease control. A practical field manual. Vol 1. Tick control.* Rome; 1984. p. 299.

8. Bowman AS, Nuttall PA. *Ticks: biology, disease and control. Parasitology.* Cambridge: Cambridge University Press; 2004.

9. De Castro JJ. Sustainable tick and tick borne disease control in livestock improvement in developing countries. *Vet Parasitol* 1997;**71**:77−97.

10. Barnett SF. *The control of ticks on livestock.* Rome: FAO Agricultural Studies Serial No. 54; 1968.

11. Sonenshine DE. *Biology of ticks*, vol. 1. New York, NY: Oxford University Press; 1991. p. 1−449.

12. Anonymous. *Animal Health Annual Report.* Trinidad and Tobago: Ministry of Agriculture, Land and Marine Resources; 2010.

13. Neumann LG. Revision de la famille des Ixodides. *Memoires de la Societezoologique de France* 1899;**12**:107−294 cited by Smith MW. Some aspects of the ecology and lifecycle of *Amblyomma cajennense* (Fabricius 1787) in Trinidad and their influence on tick control measures. *Ann Trop Med Parasitol* 1975;**69**(1):121−9.

14. Hart JH. The tick of the domestic fowl and fowl fever. *Bull Misc Inf Trinidad* 1899;**19** (Pt II):180.

15. Cooley RA, Kohls GM. *The argasidae of North America, Central America, and Cuba.* Amer. Midland Nat. Monogr 1944; No. 1. p. 152.

16. Kohls GM. *Ixodes downsi*, a new species of tick from a cave in Trinidad, British West Indies (Acarina-Ixodidae). *Proc Entomol Soc Wash* 1957;**59**:257−64.

17. Kohls GM. A new species of *Amblyomma* from iguanas in the Caribbean (Acarina: Ixodidae). *J Med Entomol* 1969;**6**:439−42.

18. Fairchild GB, Kohls GM, Tipton V. The ticks of Panama. In: Wenzel RL, Tipton VJ, editors. *Ectoparasites of Panama.* Chicago, IL: Field Museum Nat Hist; 1966. p. 167−219.

19. Aitken THG, Worth CB, Tikasingh ES. Arbovirus studies in Bush Bush Forest, Trinidad, W.I., September 1959–December 1964. III. Entomological Studies. *Am J Trop Med Hyg* 1968;**17**(2):253–68.

20. Aitken THG, Jonkers AH, Tikasingh ES, Worth CB. Hughes virus from Trinidadian ticks and terns. *J Med Entomol* 1968;**5**:501–3.

21. Kohls GM, Hoogstraal H, Clifford CM, Kaiser MN. The subgenus *Persicargas* (Ixodoidea, Argasidae, *Argas*). 9. Redescription and new world records of *Argas (P.) persicus* (Oken), and resurrection, redescription, and records of *A.(P.) radiatus* Railliet, *A.(P.) sanchezi* Duges, and *A. (P.) miniatus* Koch, new world ticks misidentified as *A. (P.) persicus. Ann Entomol Soc Am* 1970;**63**(2):590–606.

22. Everard COR, Tikasingh ES. Ecology of the rodents, *Proechimys guyannensis trinitatis* and *Oryzomys capito velutinus*, on Trinidad. *J Mammal* 1973;**54**(4):875–86.

23. Smith MW. The ecology of *Amblyomma cajennense* a parasite of man and animals in Trinidad. *Trans R Soc Trop Med Hyg* 1973;**67**(1):36.

24. Smith MW. A survey of the distribution of the Ixodid ticks *Boophilus microplus* (Canestrini, 1888) and *Amblyomma cajennense* (Fabricius, 1787) in Trinidad and Tobago and the possible influence of the survey results on planned livestock development. *Trop Agric (Trinidad)* 1974;**51**(4):559–67.

25. Rawlins SC. *Toxicological and biological studies on Jamaican and other Caribbean populations of the cattle tick* Boophilus microplus *(canestrini) (Acarina:Ixodidae) [Ph.D. thesis]*. Mona, Jamaica: The University of the West Indies; 1977.

26. Dindial P. *Studied on the life cycle and seasonal incidence of* Boophilus microplus *in Trinidad and Tobago [M.Sc. thesis]*. Saint Augustine, Trinidad: The University of the West Indies; 1977.

27. Aitken THG, Omardeen TA, Giles CD. The 1958 Cayenne tick outbreak. *J Agric Soc Trinidad* 1958;**58**:153–7.

28. Lans C. *Natural pest control*. <mhtml:file://Ticks-Caribbean/NaturalPest control.mht>; 2002.

29. Polar PA, de Muro MA, Kairo MT, Moore D, Pegram R, John S-A, et al. Thermal characteristics of *Metarhizium anisopliae* isolates important for the development of biological pesticides for the control of cattle ticks. *Vet Parasitol* 2005;**134**:159–67.

30. Polar PA. *Studied on the use of entomopathogenic fungi for the control of ticks on cattle [Ph.D. thesis]*. ST. Augustine, Trinidad: The University of the West Indies; 2007.

31. Asgarali A, Coombs DK, Mohammed F, Campbell MD, Caesar E. A serological study of *Babesia caballi* and *Theileria equi* in thoroughbreds in Trinidad. *Vet Parasitol* 2007;**144**:167–71.

32. Georges K. *Molecular diagnosis of haemopathogen infection of companion animals [Ph.D. thesis]*. St. Augustine: The University of the West Indies; 2010.

33. Poucher KL, Hutcheson HJ, Keirans JE, et al. Molecular genetic key for the identification of 17 *Ixodes* species of the United States (Acari: Ixodidae): a methods model. *J Parasitol* 1999;**85**:623–9.

34. Walker JB, Mehlitz D, Jones GE. *Notes on the Ticks of Botswana*. Eschborn, Germany, GTZ. 1978. <https://en.wikibooks.org/wiki/Special:BookSources/3880850526> ISBN 3-88085-052-6M.

35. Yunker CE, Keirans JE, Clifford CM, Easton ER. *Dermacentor* ticks (Acari: Ixodoidea: Ixodidae) of the New World: a scanning electron microscope atlas. *Proc Entomol Sot Wash* 1986;**88**:609–27.

36. Walker AR, Bouattour A, Camicas JJ, Estrada-pena A, Horak IG, Latif AA, et al. *Ticks of domestic animals in Africa A guide to identification of species*. Edinburgh, Scotland: Bioscience Reports; 2003. p. 1–122.

37. Hinton HE. The structure of the spiracles of the cattle tick, *Boophilus microplus*. *Aust J Zool* 1967;**15**(5):941−5.

38. Schol H, Dongus H, Gothe R. Morphology of spiracles in adult *Hyalomma truncatum* ticks (Acari; Ixodidae). *Exp Appl Acarol* 1995;**19**:287−306.

39. Obenchain FD, Oliver Jr. JH. The heart and arterial circulatory system of ticks (Acari: (Ixodioidea). *J Arachnol* 1976;**3**:57−74.

40. Binnington KC, Obenchain FD. Structure and function of the circulatory, nervous and neuroendocrine systems of ticks. In: Obenchain FD, Calun RL, editors. *Physiology of ticks*. Oxford: Pergamon Press; 1982. p. 351−98.

41. Balashov YS. Blood sucking ticks (Ixodoidea) vectors of disease of man and animals [English Trans.]. *Misc Publ Entomol Soc Am* 1972;**8**:161−376.

42. Anastos G, Kaufman TS, Kardarsan S. An unusual reproductive process in *Ixodes kopsteini* (Acarina: Ixodidae). *Ann Entomol Soc Am* 1973;**66**:483−4.

43. Wall RL, Shearer D. *Veterinary ectoparasites: biology, pathology and control*. Wiley Blackwell; 2001. ISBN: 978-0-632-05618-7.

44. Dipeolu OO. Research on ticks of livestock in Africa: review of the trends, advances and milestones in tick biology and ecology in the decade 1980−1989. *Int J Trop Insect Sci* 1989;**10**:723−40.

45. Kaufman WR, Lomas L. 'Male Factors' in ticks: their role in feeding and egg development. *Invertebr Reprod Dev* 1996;**30**:191−8.

46. Basu AK, Haldar DP. Biology of *Boophilus microplus* (Canestrini, 1887). *J Nat Hist (India)* 2008;**4**(2):30−4.

47. Basu AK, Mohammed A. Field and laboratory studies on the bionomics of *Ornithodoros savignyi* Audouin, 1827 in Borno State, Nigeria. In: *Proceedings of second pan commonwealth veterinary conference, Bangalore, India*; 1998. p. 1138−45.

48. Estrada-Peña A. Forecasting habitat suitability for ticks and prevention of tick-borne diseases. *Vet. Parasitol.* 2001;**98**(1-3):111−32. Available from: http://dx.doi.org/10.1016/S0304-4017(01) 00426-5.

49. Newson RW. The life cycle of *Rhipicephalus appendiculatus* on the Kenya coast. In: Wilde JKH, editor. *Tickborne diseases and their vectors, Proc. Int. Conf. Edinburgh*; 1978. p. 46−50.

50. Soulsby EJL. *Helminths, arthropods and protozoa of domesticated animals*. ELBS; 1982.

51. Martins JR, Furlong J. Avermectin resistance of the cattle tick *Boophilus microplus* in Brazil. *Vet Rec* 2001;**149**:64.

52. Romero NA, Benavides OE, Herrera GC, Parra TMH. Resistance of the tick *Boophilus microplus* to organophosphate and synthetic pyrethroid acaricides in the department of Huila. *Rev Colomb Entomol* 1997;**23**(1/2):9−17.

53. Hagen S. *A survey for acaricide resistance in the cattle tick,* Boophilus microplus *in Central America*. Hannover: Institut fur Parasitologie, Tierarztlichen Hochschule Hannover; 1997.

54. Mendes MC, Lima CKP, Nogueira AHC, et al. Resistance to cypermethrin, eltamethrin and chlorpyriphos in populations of *Rhipicephalus (Boophilus) microplus* (Acari:Ixodidae) from small farms of the State of São Paulo, Brazil. *Vet Parasitol* 2011;**178**(3−4):383−8.

55. Rodriguez-Vivas RI, Trees AJ, Rosado-Aguilar JA, Villegas-Perez SL, Hodgkinson JE. Evolution of acaricide resistance: phenotypic and genotypic changes in field populations of *Rhipicephalus (Boophilus) microplus* in response to pyrethroid selection pressure. *Int J Parasitol* 2011;**41**:895−903.

56. Kagaruki LK. Tick (Acari: Ixododae) resistance to organochlorine acaricides in Tanzania. *Trop Pest Manag* 1991;**37**(1):33−6.

57. Basu AK, Haldar DP. A note on the effect of continuous use of sevin 50 WP on some cattle ticks. *J Vet Parasitol* 1997;**11**(2):183−4.

58. Coronado A. Current status of the tropical cattle tick *Boophilus microplus* in Venezuela. In: Rodriguez C, Sergio D, Fragoso H, editors. *3rd International seminary on animal parasitology. Acapulco, Guerrero, Mexico*; 1995, vol. 11−13. p. 22−8.

59. Benavides OE. *Boophilus microplus* tick resistance to acaricides in Colombia. A summary of the present situation. In: Rodriguez C, Sergio D, Fragoso H, editors. *3rd International seminary on animal parasitology. Acapulco, Guerrero, Mexico*; 1995, vol. 11−13. p. 22−8.

60. Kumar S, Paul S, Sharma AK, et al. Diazinon resistant status in *Rhipicephalus (Boophilus) microplus* collected from different agro-climatic regions of India. *Vet Parasitol* 2011;**181**(2−4):274−81.

61. Abbas RZ, Zamanb MA, Colwell DD, Gillearde J, Iqbal Z. Acaricide resistance in cattle ticks and approaches to its management: the state of play. *Vet Parasitol* 2014;**203**:6−20.

62. Rawlins SC, Mansingh A. Patterns of resistance to various acaricides in some Jamaican populations of *Boophilus microplus*. *J Econ Entomol* 1978;**71**:956−60.

63. Allen JR, Humphreys SJ. Immunisation of guinea pigs and cattle against ticks. *Nature (London)* 1979;**280**:491−3.

64. Willadsen P. Immunity to ticks. *Adv Parasitol* 1980;**18**:293−313.

65. Allen JR, Nelson WA. Immunological responses to ectoparasites. *Fortschr Zool* 1982;**12**:169−80 Band 27 Zbl. Bakt. Suppl.

66. Johnston LAY, Kemp DH, Person RD. Immunization of cattle against *B. microplus* using extract derived from adult female ticks: effects of induced immunity on tick populations. *Int J Parasitol* 1986;**16**:27−34.

67. Kemp DH, Agbede RIS, Johnston LAY, Gough JM. Immunization of cattle against *B. microplus* using extracts derived from adult female ticks feeding and survival of the parasite on vaccinated cattle. *Int J Parasitol* 1986;**16**:115−20.

68. Willadsen P, Bird P, Cobon GS, Hungerford J. Commercialization of a recombinant vaccine against *Boophilus microplus*. *Parasitology* 1995;**110**:S43−50.

69. Almazán C, Kocan KM, Bergman DK, Garcia-Garcia JC, Blouin EF, de la Fuente J. Identification of protective antigens for the control of *Ixodes scapularis* infestations using cDNA expression library immunization. *Vaccine* 2003;**21**:1492−501.

70. de la Fuente J, Kocan KM. Advances in the identification and characterization of protective antigens for development of recombinant vaccines against tick infestations. *Exp Rev Vaccines* 2003;**2**:583−93.

71. Willadsen P. Anti-tick vaccines. *Parasitology* 2004;**129**:S367−87.

72. Sonenshine DE, Kocan KM, de la Fuente J. Tick control: further thoughts on a research agenda. *Trends Parasitol* 2006;**22**:550−1.

73. Willadsen P. Tick control: thoughts on a research agenda. *Vet Parasitol* 2006;**138**:161−8.

74. de la Fuente J, Kocan KM, Blouin EF. Tick vaccines and the transmission of tick-borne pathogens. *Vet Res Com* 2007;**31**((Suppl. 1):85−90.

75. Merino O, Alberdi P, Perez de la Lastra JM, De la Fuente J. Tick vaccines and the control of tick-borne pathogens. *Cell Infect Microbiol* 2013;**3**:30.

76. Willadsen P, Riding GA, McKenna RV, Kemp DH, Tellam RL, Nielsen JN, et al. Immunological control of a parasitic arthropod. Identification of a protective antigen from *Boophilus microplus*. *J Immunol* 1989;**143**:1346−51.

77. de la Fuente J, Rodriguez M, Redondo M, et al. Field studies and cost-effectiveness analysis of vaccination with Gavac against the cattle tick *Boophilus microplus*. *Vaccine* 1998;**16**:366−73.

78. de la Fuente J, Rodriguez M, Montero C, et al. Vaccination against ticks (*Boophilus* spp.): the experience with the Bm86-based vaccine Gavac. *Genet Anal* 1999;**15**:143−8.

79. Gómez RU. *Effect of application of GAVAC TM plus integrated tick control on acaricide treatments in Cuba*. <http://www.fao.org/fileadmin/templates/abdc/documents/ubieta.pdf>; 2010.

80. Basu AK, Basu M, Adesiyun AA. A review on ticks (Acari: Ixodoidea: Ixodidae, Argasidae), associated pathogens and diseases of Trinidad and Tobago. *Acarologia* 2012;**52** (1):39−50.

81. Cruz JOdela. In: Woods CA, Sergile FE, editors. *Biogeography of West Indies. Patterns in the biogeography of West Indian ticks*. Boca Raton, FL: CRC Press; 2001. p. 7.

82. Thompson GB. Ticks of Jamaica, BWI. Records and notes (including a summary of the distribution of the West Indian species). *Ann Mag Nat Hist* 1950;**12**(3):220−9.

83. Drummond RO, Cadogan BL, Wilson TM. Ticks on livestock in Barbados. *Veterinary Parasitol* 1981;**8**(3):253−9.

84. Rawlins SC, Mahadeo S, Martinez R. A list of the ticks affecting man and animals in the Caribbean. *Caraphin News* 1993;**6**:8−9.

85. Camus E, Barre N. Vector situation of tick-borne diseases in the Caribbean islands. *Vet Parasitol* 1995;**57**:167−76.

86. Loftis AD, Kelly PT, Freeman MD, Fitsharris S, Beeler-Marfisi J, Wang C. Tick-borne pathogens and diseases in dogs on St. Kitts, West Indies. *Vet Parasitol* 2013;**196**(1−2):44−9.

87. Myers H, DeSocio A, Bailey A, Werners-Butler C.M. *Prevalence and incidence of ticks and tick borne diseases in native Grenadian horses*. St. George's University Research Day and Phi Zeta Research Emphasis Day Saturday 16 February 2013. <www.sgu.edu/research/pdf/phizeta-2013-abstracts/Link21.pdf>.

88. Rawlins SC, Mansingh A. A review of tick and screwworms affecting livestock in the Caribbean. *Insect Sci Appl* 1987;**8**:259−67.

89. Uilenberg G, Barre N, Camus E, Burridge MJ, Garris GI. Heartwater in the Caribbean. *Prev Vet Med* 1984;**2**:255−67.

90. Barre N, Garris GI, Camus D. Propagation of the tick *Amblyomma variegatum* in the Caribbean. *Rev Sci Techn* 1995;**14**:841−55.

91. Pegram RG, De Castro JJ, Wilson DD. The CARICOM/FAO/IICA Caribbean *Amblyomma* Program. *Ann NY Acad Sci* 1998;**849**:343−8.

92. Ahoussou S, Lancelot R, Sanford B, Porphyre T, Bartlette-Powell P, Compton E, et al. Analysis of *Amblyomma* surveillance data in the Caribbean: lessons for future control programmes. *Vet Parasitol* 2010;**167**(2−4):327−35.

93. Camus E, Barre N. *Amblyomma variegatum* and associated diseases in Caribbean, strategies for control and for eradication in Guadeloupe. *Parasitologia (Rome)* 1990;**32**:185−93.

94. FAO 2013. <http://faostat.fao.org/site/573/DesktopDefault.aspx?PageID=573#ancor>.

95. Thomas RA. *Studies on the biology of the cattle tick* Boophilus microplus*, parasitic on cattle at the field station*. Trinidad: DTA Report. The University of the West Indies; 1963.

96. Amorim M, Career MCP, Serra-Freire NM. Morphological and chaetotaxy aspects of larvae of *Argas (Persicargas) miniatus* (Acari: Argasidae) in Brazil. *Braz Arch Vet Med Zootechny* 2003;**55**(4). Available from: http://dx.doi.org/10.1590/S0102-09352003000400019.

97. Guglielmone AA, Estrada-Pena A, Keirans JE, Robbins RG. Ticks (Acari. Ixodida) of the neotropical zoogeographic region. In: *Special Publication of the International Consortium on Ticks and Tick-borne Diseases*, The Netherlands; 2003. p. 174.

98. Hoffmann A, Lopez-Campo G. *Biodiversidad de los ácaros en México*. México: Jiménez Editores e Impresores; 2000 p. 230.

99. Kakarsulemankhel JK. Re-description and new record of *Argas (Persicargas) persicus* (Oken, 1881) (Acarina: Argasidae) from Balochistan, Pakistan. *Pak Entomol* 2010;**32** (No. 2):82–94.

100. Labuda M, Nuttall PA. Tick-borne viruses. *Parasitology* 2004;**129**:S221–45.

101. Estrada-Pena A, Jongejan F. Ticks feeding on humans: a review of records on human-biting Ixodoidea with special reference to pathogen transmission. *Exp Appl Acarol* 1999;**23**: 685–715.

102. Kohls GM, Sonenshine DE, Clifford CM. The systemics of the subfamily Ornithodorinae (Acarina: Argasidae). II. Identification of the larvae of the Western Hemisphere and descriptions of three new species. *Ann Entomol Soc Am* 1965;**58**(3):331–64.

103. Dantas-Torres F, Venzal JM, Bernardi LFO, Ferreira RL, Onofrio VC, Marcili A, et al. Description of a new species of bat-associated argasid tick (acari: argasidae) from Brazil. *J Parasitol* 2012;**98**(1):36–45.

104. Arthur DR. *British ticks*. London: Butterworths; 1963. 213 pp.

105. Keirans JE, Hutcheson HJ, Oliver JH. *Ornithodoros (Alectorobius) capensis* Neumann (Acari: Ixodoidea: Argasidae), a parasite of seabirds, established along the southeastern sea-coast of the United States. *J Med Entomol* 1992;**29**(2):371–3.

106. Jonkers AH, Casals, Aitken THG, Spence L. Soldado Virus, a new agent from Trinidadian *Ornithodoros* ticks. *J Med Entomol* 1973;**10**:517–19.

107. Hoogstraal H. [chapter 18] In: Gibbs AJ, editor. *Viruses and invertebrates*. Amsterdam and London: North-Holland Publ. Co.; 1973. p. 349–90.

108. Chastel C, Bailly-Choumara H, Le Lay G. Pouvoir pathogéne naturel pour l'homme d'un variant antigénique du virus Soldado isolé au Maroc. *Bull Soc Path Exot* 1981;**74**:499–505.

109. Aitken THG, Spence L, Jonkers AH, Downs WG. A 10-year survey of Trinidad arthropods for natural virus infection (1953-1963). *J Med Ent* 1969;**6**:207–15.

110. Denmark HA, Clifford CM. A tick of the *Ornithodoros capensis* group established on Bush Key, Dry Tortugas, Florida. *Florida Entomol* 1962;**45**(3):139–42.

111. Nava S, Venzal JM, Diaz MM, Mangold AJ, Guglielmone AA. The *Ornithodoros hasei* (Schulze, 1935) (Acari: Argasidae) species group in Argentina. *Syst Appl Acarol* 2007;**12**:27–30.

112. Nava S, Lareschi M, Rebollo C, Benitez Usher C, Beati L, Robbins RG, et al. The ticks (Acari: Ixodida: Argasidae, Ixodidae) of Paraguay. *Ann Trop Med Parasitol* 2007;**101** (3):255–70.

113. Bermúdez S, Miranda RJ, Cleghorn J, Venzal JM. *Ornithodoros (Alectorobius) puertoricensis* (Ixodida: Argasidae) Parasitizing exotic reptiles pets in Panama. *Rev FAVE—Sección Cienc Vet* 2015;**14**:1–5. Available from: www.dx.doi.org/10.14409/favecv.v14i1/3.5095.

114. Fox I. *Ornithodoros puertoricensis*, a new tick from rats in Puerto Rico. *J Parasitol* 1947;**33** (3):253–9.

115. Paternina LE, Díaz-Olmos Y, Paternina-Gómez M, Bejarano EE. *Canis familiaris*, a new host of *Ornithodoros (A.) puertoricensis* Fox, 1947 (Acari: Ixodida) in Colombia. *Acta Biol Colomb* 2009;**14**(1):153–60.

116. Endris RG, Haslett TM, Hess WR. African swine fever in the soft tick, *Ornithodoros (Alectorobius) puertoricensis* (Acari: Argasidae). *J Med Entomol* 1992;**29**:900–94.

117. Endris RG, Keirans JE, Robbins RG, Hess WR. *Ornithodoros (Alectorobius) puertoricensis* (Acari: Argasidae): redescription by scanning electron microscopy. *J Med Entomol* 1989;**26** (3):146–54.

118. Nava S, José M, Venzal EA, Novel RE, Mangold AJ, Labruna MB. Morphological study of *Ornithodoros viguerasi* Cooley and Kohls, 1941 (Acari: Ixodida: Argasidae), with sequence information from the mitochondrial 16S rDNA Gene. *Acarologia* 2012;**52**(1):29–38.

119. Jones EK, Clifford CM, Keirans JE, Kohls GM. The ticks of Venezuela (Acarina: Ixodoidea) with a key to the species of *Amblyomma* in the western hemisphere. *Brigham Young Univ Sci Bull Biol Ser* 1972;**17**(4):25.

120. Voltzit OV. A review of neotropical *Amblyomma* species (Acari: Ixodidae). *Acarina* 2007;**15**(1):3–134.

121. Guglielmone AA, Estrada-Pena A, Luciani CA, Mangold AJ, Keirans JE. Hosts and distribution of *Amblyomma auricularium* (Conil 1878) and *Amblyomma pseudoconcolor* Aragao, 1908 (Acari: Ixododae). *Exp Appl Acarol* 2003;**29**(1–2):131–9.

122. Nuttall GHF, Warburton C, Robinson LF. *Ticks: a monograph of the Ixodoidea. Part IV: the genus* Amblyomma. London: Cambridge University Press; 1926.

123. Clarkson MJ. *Tick-borne diseases of cattle in Trinidad and Tobago.* London: Report to the Ministry of Overseas Development; 1969.

124. Smith MW. Some aspects of the ecology and lifecycle of *Amblyomma cajennense* (Fabricius 1787) in Trinidad and their influence on tick control measures. *Ann Trop Med Parasitol* 1975;**69**(1):121–9.

125. Newstead R. Reports of the twenty-first expedition of the Liverpool School of Tropical Medicine, Jamaica, 1908-1909, Section I, Medical and Economic Entomology. *Ann Trop Med Parasitol* 1909;**3**:421–69 Pls. XIII–XV, 2 Text-figs. (In: Ticks—a monograph of the Ixodoidea by Nuttall).

126. Stoll O. Arachnida Acaridia. *Biologia Centrali-Americana*, Zool. London (In: Ticks—a monograph of the Ixodoidea by Nuttall); 1886–1893, p. v–xxi, 1–55, 21 plates.

127. Estrada-Penã A, Jongejan F. Ticks feeding on humans: a review of records on human-biting Ixodoidea with special reference to pathogen transmission. *Exp Appl Acarol* 1999;**23**:685–715.

128. Parola P, Paddock CD, Raoult D. Tick-borne rickettsioses around the world: emerging disease challenging old concepts. *Clin Microbiol Rev* 2005;**18**:719–56.

129. McDade JE, Newhouse VF. Natural history of *Rickettsia rickettsii. Annu Rev Microbiol* 1986;**40**:287–309.

130. Rawlins SC. *Toxicological and biological studies on Jamaican and other Caribbean populations of the cattle tick* Boophilus microplus *(canestrini) (Acarina:Ixodidae) [Ph.D. thesis].* Mona, Jamaica: The University of the West Indies; 1977.

131. Turk F. On two new species of tick. *Parasitology* 1948;**38**(4):243–6.

132. Keirans JE. The tick collection (Acarina: Ixodidae) of the Hon. Nathaniel Charles Rothschild deposited in the Nuttall and general collections of the British Museum (Natural History). *Bull Br Museum Nat Hist Zool Ser* 1982;**42**:1–36.

133. Robbins RG, Deem SL, Occi JL. First report of *Amblyomma humerale* Koch (Acari: Ixodida: Ixodidae) from Bolivia, with a synopsis of collections of this tick from the South American yellow-footed tortoise, *Geochelone denticulata* (L.) (Reptilia: Testudines: Testudinidae). *Proc Entomol Soc Wash* 2003;**105**:502–5.

134. Morshed MG, Scott JD, Keerthi F, Beati L, Mazerolle DF, Geddes G, et al. Migratory songbirds disperse ticks across Canada, and first isolation of the lyme disease spirochete, *Borrelia burgdorferi*, from the avian tick, *Ixodes auritulus. J Parasitol* 2005;**91**:780–90.

135. Labruna MB, Camargo LM, Terrassini FA, Schumaker TTS, Camargo EP. Notes on parasitism by *Amblyomma humerale* (Acari: Ixodidae) in the State of Rondonia, Western Amazon, Brazil. *J Med Entomol* 2002;**39**:814−17.

136. Barros-Battesti DM, Arzua M, Bechara GH. *Carrapatos de importt111Cia medicoveterinaria da regido Neotropical. Unlguia' ilustradopara identifica9iio de especies.* Sao Paulo: VoxlICTfD-3/Butantan; 2006. 223 pp.

137. Nava S, Valazco PM, Guglielmone AA. First record of *Amblyomma longirostre* (Koch, 1844) (Acari: Ixodidae) from Peru, with a review of this tick's host relationships. *Syst Appl Acarol* 2010;**15**:21−30.

138. Venzal JM, Castro O, Claramunt S, Guglielmone AA. Primer registro de *Amblyomma longirostre* (Acari: Ixodida) en Uruguay. *Parasitol Latinoamericana* 2003;**58**:72−4.

139. Labruna MB, Whitworth T, Horta MC, Bouyer DH, McBride JW, Pinter A, et al. *Rickettsia* species infecting *Amblyomma cooperi* ticks from an area in the state of Sao Paulo, Brazil, where Brazilian spotted fever is endemic. *J Clin Microbiol* 2004;**42**:90−8.

140. Ogrzewalska M, Pacheco RC, Uezu A, Richtzenhain LJ, Ferreira F, Labruna MB. Rickettsial infection in *Amblyomma nodosum* ticks (Acari:Ixodidae) from Brazil. *Ann Trop Med Parasitol* 2009;**103**(5):413−25.

141. Burridge MJ. *Non-native and invasive ticks. Threats to human and animal health in the United States.* Gainesville: University of Florida Press; 2011. 448 pp.

142. Forlano M, Scofield A, Elisei C, Fernandes KR, et al. Diagnosis of *Hepatozoon* spp. in *Amblyomma ovale* and its experimental transmission in domestic dogs in Brazil. *Vet Parasitol* 2005;**134**:1−7.

143. Guglielmone AA, Nava S. Hosts of *Amblyomma dissi*mile Koch, 1844 and *Amblyomma rotundatum* Koch, 1844 (Acari: Ixodidae). *Zootaxa* 2010;**2541**:27−49.

144. Floch H, Fauman P. Ixodides de la Guyane et des Autilles Francaises. *Arch l'inst Pasteur Guyane Fr Publ* 1958;**19**(No. 446):1−94.

145. Guglielmone AA, Beati L, Barros-Battesti DM, Labruna MB, Nava S, Venzal JM, et al. Ticks (Ixodoidea) on humans in South America. *Exp Appl Acarol* 2006;**40**:83−100.

146. Guglielmone AA, Robbins RG, Apanaskevich DA, Petney TN, Estrada-Peña A, Horak IG. *The hard ticks of the world.* Hiedelberg: Springer; 2014.

147. Cooley RA. *The genera* Boophilus Rhipicephalus *and* Haemaphysalis *(Ixodidae) of the new Worlds.* Washington: National Institute of Health Bulletin No.187, United States, Government Printing Office; 1946 p. 1−55.

148. Kohls GM. Records and new synonymy of new world *Haemaphysalis* ticks, with descriptions of the nymph and larva of *H. juxtakochi* Cooley. *J Parasitol* 1960;**46**:355−61.

149. Valeria ND, Nava S, Cirignoli S, Guglielmone AA, Poi ASG. Ticks (Acari: Ixodidae) parasitizing endemic and exotic wild mammals in the Esteros del Iberá wetlands. *Argentina Syst Appl Acarol* 2012;**17**(3):243−50.

150. Labruna MB, Pacheco RC, Richtzenhain LJ, Szabo MPJ. Isolation of *Rickettsia rhipicephali* and *Rickettsia bellii* from *Haemaphysalis juxtak*ochi Ticks in the State of São Paulo, Brazil. *Appl Environ Microbiol* 2007;**73**(3):869−73.

151. Gonzalez-Acuna D, Nava S, Mangold AJ, Guglielmone AA. *Ixodes downsi* Kohls, 1957 in Venezuela. *Syst Appl Acarol* 2008;**13**:39−42.

152. Wilson N, Baker WW. *Ixodes downsi* (Acari: Ixodidae) from Peru. *Proc Entomol Soc Wash* 1988;**91**:54.

153. Guzmán-Cornejo C, Robbins RG. The Genus *Ixodes* (Acari: Ixodidae) in Mexico: adult identification keys, diagnoses, hosts, and distribution. *Rev Mex Biodiversidad* 2010;**81**:289−98.

154. Luz HR, Faccini JL, Landulfo GA, Sampaio Jdos S, Costa Neto SF, Famadas KM, et al. New host records of *Ixodes luciae* (Acari: Ixodidae) in the State of Para, Brazil. *Rev Bras Parasitol Vet* 2013;**22**(1):152—4.

155. Díaz MM, Nava S, Guglielmone AA. The parasitism of *Ixodes luciae* (Acari: Ixodidae) on marsupials and rodents in Peruvian Amazon. *Acta Amaz* 2009;**39**(4):997—1002. Available from: http://dx.doi.org/10.1590/S0044-59672009000400029.

156. Guglielmone AA, Nava AA. Rodents of the subfamily Sigmodontinae (Myomorpha: Cricetidae) as hosts for South American hard ticks (Acari: Ixodidae) with hypotheses on life history. *Zootaxa* 2011;**2904**:45—65.

157. de la Cruz J, Socarras AA, García MJ. Acari causing zoonoses in Cuba. *Rev Cubana Cienc Vet* 1991;**22**:101—5.

158. Guglielmone AA, Mangold AJ, Vinabal AE. Ticks (Ixodidae) parasitizing humans in four provinces of north-western Argentina. *Ann Trop Med Parasitol* 1991;**85**:539—42.

159. Kuo-Fan T. Acari: Ixodidae. *Econ Insect Fauna China* 1991;**39**:1—359.

160. Williams HE, Gonzalez FO. Two tick-borne diseases affecting exotic cattle introduce into Trinidad. *Trop Agric Trinidad* 1968;**45**:22—32.

161. Dantas-Torres F. The brown dog tick, *Rhipicephalus sanguineus* (Latreille, 1806) (Acari: Ixodidae): from taxonomy to control. *Vet Parasitol* 2008;**152**:173—85.

162. Dantas-Torres F, Figueredo LA, Brandão-Filho SP. *Rhipicephalus sanguineus* (Acari: Ixodidae), the brown dog tick, parasitizing humans in Brazil. *Rev Soc Brasil Med Trop* 2006;**39**(1):64—7.

163. Rawlins SC, Tikasingh ES, Martinez R. *Catalogue of haematophagous arthropods in the Caribbean region*. Unpublished Report. Trinidad: Caribbean Epidemiology Centre (CAREC); 1992. 168 pp.

164. Alderink FJ, McCauley EH. The probability of the spread of *Amblyomma variegatum* in the Caribbean. *Prev Vet Med* 1988;**6**:285—94.

165. Uilenberg G. Acquisitions nouvellesdans la connaissance du role vecteur de genre *Amblyomma* (Ixodidae). *Rev. Elev. Med. Vet. Pays. Trop.* 1983;**36**(1):61—6.

166. Jongejan F. Studies on the transmission of *Cowdria ruminantium* by African and American *Amblyomma* ticks. In: Munderloh UG, Kurti TJ, editors. *Proceedings and abstracts, 1st Int. conf. on tick-borne pathogens at the Host-Vector Interface: an agenda for research, 15—18 Sept. St. Paul, MN*; 1992. p. 143.

167. Hunter S. *The cattle tick Boophilus annulatus: studies of the effect on the dairy cow in Trinidad [A.I.C.T.A. thesis]*. Trinidad: The University of the West Indies; 1945.

Printed in the United States
By Bookmasters